Petroleum Development and Environmental Conflict in Aotearoa New Zealand

Petroleum Development and Environmental Conflict in Aotearoa New Zealand

Texas of the South Pacific

Terrence M. Loomis

LEXINGTON BOOKS
Lanham • Boulder • New York • London

Published by Lexington Books
An imprint of The Rowman & Littlefield Publishing Group, Inc.
4501 Forbes Boulevard, Suite 200, Lanham, Maryland 20706
www.rowman.com

Unit A, Whitacre Mews, 26-34 Stannary Street, London SE11 4AB

British Library Cataloguing in Publication Information Available

Library of Congress Cataloging-in-Publication Data

Names: Loomis, Terrence, author.
Title: Petroleum development and environmental conflict in Aotearoa/New Zealand :
   Texas of the South Pacific / Terrence M. Loomis.
Description: Lanham, Maryland : Lexington Books, 2017. | Includes bibliographical
   references and index.
Identifiers: LCCN 2016043221 (print) | LCCN 2016048573 (ebook) | ISBN
   9781498537575 (cloth : alk. paper) | ISBN 9781498537582 (Electronic)
Subjects: LCSH: Petroleum industry and trade—New Zealand. | Petroleum industry
   and trade—Environmental aspects—New Zealand. | Petroleum industry and
   trade—Government policy—New Zealand. | Environmentalism—New Zealand. |
   Environmental protection—New Zealand.
Classification: LCC HD9578.N45 L66 2017 (print) | LCC HD9578.N45 (ebook) | DDC
   333.8/23150993—dc23
LC record available at https://lccn.loc.gov/2016043221

Printed in the United States of America

Contents

Figures and Tables

Acknowledgments

Firstly, I need to explain the use of the name "Aotearoa New Zealand" in the title of this book. Aotearoa is the Māori name for New Zealand. It is often translated "Land of the long white cloud." Māori is recognized as an official language of New Zealand, and it has become the convention in recent years to cite both the Māori and English names when referring to significant geographic locations, historic sites, government departments, or educational institutions. Many people use the name "Aotearoa New Zealand" in recognition of the country's bicultural heritage and Māori sovereignty as enshrined in the Treaty of Waitangi of 1840. I share this view, and use "New Zealand" throughout the book as shorthand for Aotearoa New Zealand.

A number of people have given of their time and provided assistance with my research and in the preparation of the manuscript. I must thank all those named and confidential interviewees who shared their experiences, thoughts and information with me. I am also greatly indebted to Dr. Michael Macaulay and the Institute of Governance and Policy Studies at Victoria University of Wellington for offering me a position as a visiting research scholar and providing me with office space and administrative support during my fieldwork. The staff at Lexington Press, notably editor Amy King and assistant editors Kasey Beduhn and James Hamill, have provided invaluable guidance and support in preparation of my manuscript. I should also express my appreciation to the independent reviewer who gave valuable feedback and made many insightful suggestions on how I could improve the analysis and communicate my findings better to the reader.

I would like to express my appreciation to Kiwa Mihaka of Rongowhakaata, Ngai Tāmanuhiri, and Te Aitanga a Māhaki iwi for allowing the image of the pou whenua (territorial boundary post) which he carved to be used for the cover of this book. The pou whenua marks the ancient Pa (fortified village site) of Ngai Tāmanuhiri iwi on Te Kuri/Young Nicks Head overlooking Poverty Bay and Gisborne. This headland was sighted by Captain Cook in 1769 on the East Coast of the North Island of New Zealand.

Finally and most of all, I am indebted to my wife, artist Jean E. Loomis, for creating the art piece for the cover, for her assistance in proofreading a draft of my manuscript, and for her patience, forbearance, and loving support during the long journey to bring this research project to fruition.

List of Abbreviations

CJT	Climate Justice Taranaki
COGA	Colorado Oil and Gas Association
COP21	The 2015 United Nations Paris Climate Conference (aka Conference of Parties 21)
CSR	Corporate Social Responsibility
DOC	Department of Conservation
E&P	Exploration and production (or "upstream") companies
ECO	Environmental and Conservation Organisations of Aotearoa New Zealand
EDS	Environmental Defence Society
EEZ	Exclusive Economic Zone Act and Continental Shelf (Environmental Effects) Act 2012
EIA	Energy Information Agency (US)
EIS	Environmental Impact Statement
EROEI	Energy Returned on Energy Invested
FFT	Frack-Free Tairāwhiti
GNS Science	Geological and Nuclear Sciences (a New Zealand Crown Research Institute)
GDC	Gisborne District Council (a unitary authority, with district and regional council functions)
IEA	International Energy Agency
IPCC	Intergovernmental Panel on Climate Change
JV	Joint Venture
LGA	Local Government Act 2002
LGNZ	Local Government New Zealand
MBIE	Ministry of Business, Innovation and Employment
MfE	Ministry for the Environment

MOU	Memorandum of Understanding
MSM	mainstream media
NAK	Ngā Ariki Kaipūtahi iwi
NZEC	New Zealand Energy Corporation
NZOG	New Zealand Oil and Gas
NZPaM	New Zealand Petroleum and Minerals
NZPC	The New Zealand Petroleum Club
NZTE	New Zealand Trade and Enterprise
ODT	*Otago Daily Times*
OPEC	Organization of Petroleum Exporting Countries
PCE	New Zealand Parliamentary Commissioner for the Environment
PEPANZ	Petroleum Exploration and Production Association of New Zealand
RMA	Resource Management Act 1991
STOS	Shell Todd Oil Services
Straterra	Natural Resources of New Zealand (a mining association)
TEW	Taranaki Energy Watch
TPPA	Trans-Pacific Partnership Agreement
TRC	Taranaki Regional Council
WBCSD	World Business Council for Sustainable Development

Introduction

Since the turn of the millennium, a worldwide boom has occurred in what the petroleum industry calls "unconventional" oil and gas development, particularly directional drilling and hydraulic fracturing (or fracking). To date the boom has been centered in North America, though countries like Argentina, Mexico, Chile, Columbia, and Indonesia have recently begun getting in on the act. In the US unconventional shale gas and "tight" oil now account for 50 percent of all petroleum produced. In Canada tar sands are the leading form of unconventional production, while in Australia it's about coal seam gas (CSG) extraction. The International Energy Agency and petroleum giants like BP and ExxonMobil expect that by mid-century 35 to 40 percent of the world's energy demand will be met from unconventional sources. Whether the boom ultimately delivers the benefits that oil and gas promoters claim, given mounting evidence of the externalities (harm) the industry creates and concerns over climate change, is another matter.

"Frontier" petroleum producers like New Zealand are hoping to cash in on the unconventional boom onshore as well as from offshore sources. Shortly after the National Party came to power in 2008, it set out a policy framework called the Business Growth Agenda which called for a major expansion of the oil and gas industry. The new government under Prime Minister John Key declared it was hoping for a "game changing" discovery to transform the economy and boost export-led growth. To achieve its aims, National undertook a number of orchestrated steps in close cooperation with the petroleum industry to remove perceived impediments to oil and gas expansion, promote the industry to international investors as well as "Middle New Zealand," and co-opt or marginalize environmental opposition. The petroleum industry developed its own set of strategies or borrowed them from overseas to help achieve their mutual aims. New Zealand's experience provides a useful case study with which to address several gaps in the literature on social conflict, economic development, and the extractive industry. This study examines some of these government maneuvers and oil industry strategies more closely. It also explores how resistance and counter-strategies by environmental organizations, local anti-fracking groups, and Māori activists disrupted government/industry efforts and altered institutional relations between the state, Big Oil, and the community sector. These are the key issues that guide this study.

Although the following account encompasses political processes, petroleum industry strategies, and environmental movements at a national level, we will examine developments on the East Coast in more detail later in the book. The East Coast is one of the regions the government and industry have targeted for potential expansion of unconventional and offshore exploration outside the country's existing oil region of Taranaki, and where they have concentrated their efforts at winning over the public and undercutting local opposition. It is also the region where I lived for a considerable portion of this study, and where I undertook much of my research.

Chapter 1 begins with a discussion of the political economy of oil and gas on a global scale. At a national level, to account for the structural effects of the struggle between the government/petroleum industry "partnership" and environmental and community anti-oil movements, the study adopts a Regulation Theory approach. From this perspective the social system of production, distribution, and accumulation can be conceived of as a triangle of contending or collaborative relationships between the private, public, and civic sectors (or more specifically the powerful interest groups within them) which change over time and shape the nature of institutions, the norms and values that guide their functioning and interactions, and the types of policies and regulatory regimes that are developed. The chapter looks at examples of countries that have succumbed to the "resource curse" versus those that have avoided the trap, and considers the implications for where New Zealand might be heading. The chapter concludes with a discussion of the typical maneuvers governments have adopted and the strategies oil companies have utilized, often in cooperation with one another, to promote and defend the industry, intimidate or hoodwink communities, and subvert environmental opposition.

Chapter 2 describes the world-wide boom in unconventional oil and gas exploration and production, questions whether the boom is all it's cracked up to be, and examines evidence of the externalities (harm) the industry creates and tries to cover up in order to remain profitable and retain its "social license to operate."

Chapter 3 begins with a brief account of the political economy of oil and gas in New Zealand. It then documents how the National Government collaborated with the oil and gas industry to systematically dismantle legislation, policies and programs put in place by previous governments that, particularly in the case of the Resource Management Act, maintained at least a tenuous balance between sustainability and economic growth. The National Government replaced this "balanced" approach with a neoliberal framework called its Business Growth Agenda, aimed at dramatically expanding primary exports such as oil and gas by 2025. Some critics labeled the new policy framework "development at all costs." The rest of the chapter looks at how John Key's government used promotional marketing, ministerial cheerleading and jawboning, "official" information dissemination, and agenda-driven funding (or funding cuts) to support expansion of extractive industries. The chapter notes examples of how ministers and government departments collaborated in some of these promotional activities with the very industry that they were charged with regulating and holding accountable.

Chapter 4 discusses the main strategies the New Zealand oil and gas industry employed to promote and defend itself. It considers instances where the industry adopted practices and tricks borrowed from North America to deal with regulators, the public, and prospective shareholders. The chapter also looks at how the industry utilized front organizations like the Petroleum Exploration and Production Association of New Zealand (PEPANZ) to coordinate industry strategies and communications, answer public concerns and criticism, share methods for "engaging" and "consulting" with Māori and local communities, and partner with politicians and officials to clear legislative impediments and reduce red tape claimed to be inhibiting the industry's business operations.

The Taranaki region is home to New Zealand's oil and gas industry, both onshore and offshore. The National government has sought to attract foreign investment and expand the petroleum industry through a marketing program and introduction of an annual auction of exploration blocks. While these efforts are directed at an international audience, the government also embarked on a campaign to promote the industry to the general public and "win the hearts and minds" of residents in geologically promising regions. Chapter 5 focuses on the East Coast, looking briefly at industry estimates of how much petroleum there could be, and initial exploration activities to try to discover that resource. It pays particular attention to government and industry efforts to "sell" the people of the East Coast on the idea of establishing an oil and gas industry. The chapter discusses whether a "boom" is possible or likely, the weaknesses in "factual" government reports about the prospectivity of the region, and the potential impacts an expanded oil and gas industry could have on local communities, tangata whenua (traditional guardians of the land), and on the environment.

Chapter 6 uses the boomtown model to compare experiences of communities confronting oil and gas developments in other countries with what groups and communities are doing in New Zealand. The chapter discusses examples of how local councils, tangata whenua, and community groups have responded to unconventional oil and gas developments in their area, and attempts to identify typical actions that communities and activist organizations have found useful in engaging with the state and oil companies. It concludes with brief case studies of how anti-petroleum and environmental groups in Taranaki and the East Coast have employed some of these methods in resisting the expansion of the industry.

To be clear, this is an account of how New Zealand's National government and the petroleum industry have cooperated to manipulate the legislative, regulatory, policy-making, and public relations system in order to achieve their aim of promoting and justifying the expansion of an oil and gas industry "in the national interest." Whether these efforts have been truly in the national interest, given the externalities generated by this sunset industry and rising concerns over climate change, history will judge and New Zealand citizens will ultimately determine at the ballot box.

Chapter One

The Political Economy of Oil and Gas Development

Governments live a precarious existence at the best of times, not simply because of the fickleness of voters. They must engage in a constant balancing act between promoting economic development and attracting investment while at the same time facilitating and regulating the market, monitoring business practices, punishing corporate misdeeds, and collecting tax revenues to assist those who have been directly or indirectly disadvantaged by corporate activities. Nevertheless, the agendas of governments and corporations occasionally overlap, providing a basis for acting jointly, ostensibly in the national interest. In practice governments or, more often, specific factions within them, are influenced and frequently dominated by powerful corporate interests. It is these relationships as they take on institutional form, rather than the free market on its own, that shape a country's development trajectory and determine what role if any the oil and gas industry will play in that country's development.

Civil society has the potential to influence corporate/state relations, particularly when a populist state looks to bolster its power over private corporations or when public reaction to an issue or development becomes so intense and polarized that it results in social conflict. Nowhere is this more apparent than with respect to countries where extractive industries play a major role in the economy. In his "Afterword" to Anthony Bebbington's compilation[1] of South American case studies, Stuart Kirsch suggests that in the study of how social conflict over resource exploitation influences relations between state, corporate, and community sector institutions, "there is a need for more careful scrutiny of corporate actors," particularly the strategies transnational corporations use to defend themselves and avoid accountability for the externalized costs of their operations as well as how they influence state policies and responses to protest action.[2] In addition to (1) scrutinizing petroleum corporation practices, this study examines (2) the maneuvers the New Zealand government has adopted to facilitate extractive development and deal with critics of government policy; and (3) how protest action by community groups and environmental organizations has affected government policies and resulted in a reshaping of institutional relations between extractive corporations, the states, and the community sector. These are the *three key issues* that guide the following study.

New Zealand has a tradition of being an open, liberal democracy with active citizen involvement in debating issues and local development planning. It is also at a relatively early stage in the development of unconventional oil and gas, led by a government committed to expanding the extractive sector to boost export-led growth. A few major oil companies and a handful of "juniors" have been exploring on- and offshore, and a modest amount of unconventional production has occurred in the existing oil region of Taranaki. But exploration has been slowed by the downturn in the global petroleum market and environmental/indigenous anti-oil activism. We might therefore expect, compared with countries where the unconventional boom is already a large part of the economy and generated fierce debate, that instead of a pitched battle, government actions and oil industry strategies in New Zealand would be oriented toward achieving popular acceptance, laying the groundwork for expansion, and establishing "mutually respectful" relations with municipal councils, environmentalists and community anti-oil groups. We'll see later in this study whether those expectations are borne out.

First, it's important to consider New Zealand's tip toward petroleum-backed growth in a wider global context. In this chapter we will examine how globalization has extended the reach and influence of transnational corporations in sectors like petroleum, and how this has shaped the way states engage with such industries. Then, based on a review of literature, we will attempt to identify the typical "portfolio" of strategies that petroleum corporations have utilized to promote the industry and defend themselves from criticism and attack. In a later chapter we will examine which of these strategies have been employed by the New Zealand petroleum industry in collaboration with the state, and why.

GLOBALIZATION AND THE RISING INFLUENCE
OF PETROLEUM CORPORATIONS

The oil and gas industry is a huge, complex, and constantly evolving part of the world economy upon which other sectors and governments are critically dependent to produce and distribute goods and services for their customers and citizens. Dominating this vast complex are a relatively few mega-corporations which undertake or contract out exploration, drilling, infrastructure building, production, and distribution of petroleum. Under them are an extensive network of junior exploration and production (E & P) companies and specialist field services providers, most of whom are themselves multi-million-dollar operations. The sector as a whole generates vast wealth and profits for itself and its shareholders as well as royalties and tax revenues for governments. In addition to the sector's sheer wealth, the process of globalization has increased the influence of Big Oil immensely vis-à-vis national governments.

Globalization has been defined in many ways, often too technical for the average person to understand. Put simply, globalization is the increased mobility of all forms of capital—finance, plant and equipment, people, supplies, and resources. For instance, a company that has difficulty accessing reasonably priced domestic financing for a project might obtain it offshore. Wages of local workers may be too high for a manufacturer to be competitive in the global market, so the company hires migrant workers or shifts its production to a low-wage economy. In one form or another globalization has been taking place since the dawn of the Age of Exploration. Until the twentieth century, it was driven if not always implemented by public entities—city-states, nations, and international alliances. More recently, the multinational corporation has come to the fore—expanding their economic empires, building huge portfolios of assets, and pushing for increased access to a country's land and natural resources and deregulation in order to accelerate their growth. Karliner refers to the contemporary era of globalization rather ominously as the "corporate encirclement of the planet."[3]

The globalization process has fundamentally transformed the role and authority of the nation-state, as recent debates over the Trans-Pacific Partnership Agreement (TPPA) bear witness. Governments are less able to negotiate favorable terms in international trade agreements, protect their domestic economies, impose taxes and resource royalties that would adequately compensate countries for the economic and environmental impacts of corporate operations, and enforce sanctions that would tightly regulate corporate activities. To compete with other nations, governments find themselves forced to "collaborate" with dominant sectors such as the petroleum industry that promise tax revenues, prosperity and energy security.

Korten identifies two forces, one domestic and one external, that undermine a government's ability to act in its public's best interest.[4] On the domestic front, most governments have accepted some neoliberal version of the growth paradigm as the organizing framework for their economic development policy. The domestic process is frequently referred to, as in the case of the world-renowned New Zealand Experiment which began in the mid-1980s, as "structural adjustment." As Kelsey notes,[5] neoliberal structural adjustment is about reducing the role of the state in the economy and empowering capital, through provision of incentives and removal of regulatory "impediments," to regulate itself in the marketplace:

> The "fundamentals" of the New Zealand experiment—a deregulated labour market, a minimalist government, a strict monetarist policy, the liberalisation of trade, investment and markets, and fiscal restraint—comprise an ideologically coherent package that is premised on unfettered market forces and a limited state. Their collective impact on the workplace, the public service, core social services, business and the financial system has been profound.

All of this requires a broader adjustment, enjoined by the state, of popular expectations:

> Citizens and firms are encouraged to lower their expectations and make individual, non-political adjustments to their more difficult circumstances. This may involve families sending more of their members into the workforce, selecting which children will receive a costly education, and reducing consumption.[6]

Such structural adjustment campaigns and "opening" the economy to global markets and corporate investment have resulted in increasing inequality and social malaise, people forced into part-time work or the informal economy, a deterioration of local ecosystems, and accelerated climate change. Since states are incapable of resolving these systemic problems by means of the growth paradigm (growth causes the very problems nation-states are constantly called upon to address) and with increased domination by transnational corporations, they suffer *a crisis of legitimacy* and increasing public disillusionment.[7]

Externally, nations accede to (in the case of New Zealand) or are steadily overwhelmed by "a convergence of ideological, political and technological forces behind a process of *economic globalization* that is shifting power away from governments who are responsible for the public good toward a global coterie of corporations and financial institutions driven by a single imperative, the quest for short-term financial gain."[8] Korten argues this is not some nebulous evolutionary process. Big corporations drive globalization by constructing a unified global market, avoiding citizenship in a single nation (particularly for tax purposes), influencing international trade agreements in their favor and creating a global "mono-culture of consumer preferences."[9] Because of their enormous economic influence, they are able to dominate most aspects of political life and undermine the democratic process in countries where they operate.[10] As Kelsey observes based on New Zealand experience:

When a country is vulnerable to the actions of foreign corporations, governments or financiers, the state's legal sovereignty remains intact, but the demands of these external forces can shape the direction of policy and law. Areas of influence extend beyond economic matters to decisions affecting indigenous rights, the environment, labour, social policy and human rights. Appeasing powerful international players and 'the markets' may assume greater importance for a government than the democratic choices and economic, social and cultural needs of its citizens.[11]

In some mineral-rich third-world countries this had created a phenomenon economists call the "resource curse" where extractive industries effectively control both the state and the economy. In addition to influencing political and economic institutions, they have become adept at shaping public opinion, co-opting NGOs, and manipulating public discourse in their favor.

Of course, the influence of multinational corporations shouldn't be overstated. Kelsey argues that markets are not self-regulating: "Capital needs a legal framework that sets and polices the rules for the markets in which it operates. It also requires a sound physical, technological and intellectual infrastructure, which markets do not reliably provide."[12] Similarly Giddens[13] maintains that governments, particularly where they act in concert with one another, can still exercise considerable power over transnational corporations. In addition to international conventions, they control territory, pass legislation, and exert military power which corporations do not (at least not directly). That said, Giddens's comments seem overly optimistic given the manner in which petroleum companies have been able to buy up rights to resources, control vast territories, influence or circumvent national legislation, and bankroll military conflicts like the two Iraq wars.

Supply and control of energy has become an increasingly critical issue for the global economy, particularly in the era of climate change. Larger oil and gas companies have moved aggressively to extend their ownership of petroleum resources, infrastructure, production, and transportation infrastructure around the world. Henderson[14] notes they have also extended their reach into other energy and commodity sectors (copper, bauxite, timber) and geographically into untapped areas of the globe (including New Zealand). He maintains the largest conglomerates—Royal Dutch/Shell, Exxon Mobil, Chevron Texaco, and BP Amoco—are on course to control "virtually all energy sources on the planet." For practical reasons as well as for PR purposes, they refer to themselves more and more as "the energy industry." They have entered into joint ventures (JVs) with one another to pressure states for greater access to resource extraction opportunities, justify drilling and continued reliance on fossil fuels to the public, and defend their operations from criticism and disruption by anti-oil groups.

EXPEDITING PETROLEUM-LED GROWTH:
CORPORATE AND STATE STRATEGIES

The anthropological study of economic development, power, and inequality has tended to concentrate on the role of the state and the reactions of the disadvantaged. The role of corporations in local politics and economies may be described, but precisely *how* they exercise their power and influence is less well explored. Benson and Kirsch[15] argue for a reorientation of the study of power to focus on "how corporations operate, how they engage with states and publics, and how human health and environmental problems are negotiated." They propose concentrating on the strategies that corporations employ to defend themselves from criticism and threat, "and how this facilitates the perpetuation of harm"[16] upon which their production and profits are predicated. They identify three phases of strategic response that corporations typically move between in order to deflect threats to their continued operation: problem denial, acknowledgment and token accommodation, and strategic engagement.

A closer examination of corporate strategies and tactics is certainly warranted to understand how extractive industries exert influence over states, environmental movements, local communities, and indigenous peoples, as several recent studies demonstrate.[17] Most corporations have teams of strategy specialists whose work, needless to say, is not entirely about business operations. Of course corporate strategizing takes place in an overlapping political and economic space shared with states and civil society, and prevailing circumstances at any given time affect all three sectors albeit not always in the same way or at the same time. Nevertheless it is not unusual, particularly in democratically strong states with an engaged citizenry where corporate influence and a "social license to operate" must be constantly renegotiated, that corporations seek to collaborate strategically with the ruling government rather than simply pressure them into acceding to industry demands.

The notion of "strategy" needs to be treated with caution when analyzing corporate practice and government collaboration. The term itself conveys a sense of pre-meditated planning before an action or campaign which, if interpreted strictly, can overlook more ad hoc practices. It is sometimes useful to distinguish between strategies and what might be termed "standard maneuvers." Standard maneuvers are what Bourdieu referred to as "spontaneous improvisations"[18]—actions or language that get repeated, embellished, and perhaps become encoded in institutional practice because they produce the results a corporate executive, PR manager or politician is seeking. A minister might decide to establish an independent (or independent-appearing) task force of eminent persons to consider a controversial issue rather than appoint an officials' working group, and the government ends up being able to distance itself more readily from politically unpalatable proposals. Or a corporate executive might mention "partnership" with government, being a "good corporate citizen," or being an "environmentally responsible" company in various speeches and media interviews, and the company's public standing improves in opinion surveys. Both in their way are inadvertent PR maneuvers that end up achieving desired results, and so they may get repeated as new occasions arise.

There is merit, as Benson and Kirsch suggest, in breaking down corporate strategies into overlapping phases because business imperatives lead corporations to ramp up their promotional efforts and defensive responses as the level of public criticism and protest (and the effect of these on the company's bottom line) rises. A phased approach provides a rationale for understanding which kinds of strategies are likely to be applied for particular purposes. However, they have to do primarily with petroleum industry responses to exposure of harmful practices, criticism, and attack. The industry also employs different strategies for other purposes such as enhancing their public image, influencing state policy, and strengthening relations with communities and indigenous nations. It is therefore useful to augment Benson and Kirsch's phases of defensive strategizing with a discussion of the broader repertoire of strategies and tactics corporations may choose from at any given time in response to changing political and social circumstances.

A PORTFOLIO OF PETROLEUM INDUSTRY STRATEGIES TO PROMOTE OIL AND GAS DEVELOPMENT

Based on a review of recent literature about extractive industry strategies, particularly the unconventional oil and gas industry in the United States, Canada, the United Kingdom, South America, and Australia, it is possible to identify a "portfolio" of strategies and associated tactics which corporations employ as circumstances warrant (figure 1.1). These strategies are not selected in any particular order, and several may be used at any given time. For instance, a petroleum trade association might work with government officials to develop industry-friendly regulations, while organizing a series of public meetings and media interviews to boost popular support at the same time it is launching a social media campaign to respond to the criticisms of environmentalists.

There is considerable variation in how these strategies are implemented, but the literature suggests certain tactics are often associated with particular strategies.

Figure 1.1. Petroleum Industry Strategies for Promoting Oil and Gas Development.

Influencing and Collaborating with Governments

For more than a century, petroleum companies have been able to accumulate immense economic power by "tightening their control over markets and technology through mergers, acquisitions and strategic alliances."[19] This power as we have seen has enabled them to dominate the political life of many nations, aided by the process of globalization. Even so-called junior E&P companies can wield surprising influence, particularly when as in New Zealand the government is firmly committed to expanding extractive industries as an essential component of its economic development policy.

The inclination toward corporate/government collaboration in preference to confrontation in most petroleum-producing countries is reinforced by a taken-for-granted growth paradigm and a shared neoliberal ideology which holds that a key role of the state is to facilitate infrastructure development and minimize constraints on the free market. Collaboration operates "more like any networking or shared culture-building process out of which alliances among individuals and groups emerge and evolve. There is no conspiracy, though in practical terms the consequences are much as if there were."[20] The industry strategy of influencing and collaborating with governments is typically pursued through a standard tactics such as (a) patronage and lobbying, (b) helping develop industry-friendly policies and regulations, (c) revolving door arrangements, and (d) externalizing costs.

The oil and gas industry devotes considerable time and money to lobbying and supporting friendly politicians though the use of paid lobbyists, old-boy networks, industry advocacy groups, and industry-friendly foundations. The focus is on blocking policies that could restrict corporate activities and/or impose additional costs like more stringent environmental legislation. For instance, when the state of Pennsylvania moved to impose a severance tax on wells that were reaching the end of their productive life, the oil and gas industry mounted a vigorous media and lobbying campaign to block the legislation.[21]

In 2010–2011 Australian extractive industries ran a $22 million public relations and lobbying campaign to modify a federal Resource Super Profits Tax which would have generated $200 billion in taxes over the coming decade. Instead, a Mineral Resource Rent Tax was passed that was only predicted to bring in $38.5 billion in taxes. Bottom line: The alternative legislation resulted in a $160 billion windfall for the petroleum and mining industries[22]

In the United States, industries and corporations exert considerable influence over Congress, the White House, and government agencies through patronage, lobbying, and appointments to boards and key administrative positions. During the term of the 113th Congress (2013–2014), the petroleum industry spent over US$285 million on lobbying. [23] In return the fossil fuel sector received $35 billion in exploration and production subsidies, or what some critics call "corporate welfare."[24] As industry watchdog group Oil Change International points out, that's a 10,200 percent rate of return on investment. The corporate largess is targeted particularly at politicians in petroleum-rich states. Zeese and Flowers found that the oil and gas industry made over $8 million in campaign contributions to Pennsylvania politicians between 2000 and 2012.[25]

In the United Kingdom, Foreign Office documents released to the *Guardian* under Britain's Official Information Act revealed that government policy "barely distinguishes between the British national interest and the commercial interests of its main oil and gas companies."[26] A study by the World Development Movement found that a third of UK government ministers had direct links to fossil fuel corporations and the banks that finance them. Internationally, petroleum industry lobbyists are unusually successful in getting time with politicians in countries where they do business and exercise considerable influence over policy-making.[27] The European Union recently abandoned or weakened proposals for regulating unconventional production and addressing climate change after receiving a letter from a top BP executive warning of a petroleum industry exodus from Europe if the proposals went ahead.[28]

Big oil corporations frequently employ specialist staff to keep tabs on different countries' environmental and safety regulations, challenge new rules, persuade regulators to moderate their enforcement processes, and respond to critical findings about company practices. These people often have a legal, policy or regulatory background themselves. Oil company executives may be invited to sit on government advisory boards or consultative committees, and provide feedback on policy papers or proposed legislation outside the formal public submissions process. Occasionally petroleum industry representatives or consultants are contracted to help draft policy or regulations when industry expertise is required, though governments are reluctant to publically acknowledge such involvement.[29] Of course, industry experts bring their own agendas to this kind of work besides the national interest. Karliner[30] notes that big US corporations "helped write and promote laws that use a 'risk assessment' formula to make economic considerations the determining factor over health protection when setting environmental standards, thus undermining government's ability to regulate corporate activities in the public interest."

A recent study of the US oil and gas sector discovered numerous instances of direct industry involvement in crafting regulations and legislation. For example, the Illinois Hydraulic Fracking Regulation Act "has a section on the disclosure of fracking chemicals that uses language written by Exxon and promoted by the American Legislative Exchange Council which has turned up in other state legislatures. The language would allow disclosure to be overseen by a front group for the oil and gas industry."[31] The Act was endorsed by local chapters of the Sierra Club, who cooperated in writing the regulations. Apparently they were willing to make trade-offs with the industry in order to get the language they wanted included.

Critics of the oil and gas industry have pointed to the cozy relationships that often exist between corporations and regulators. In countries with established petroleum industries, this can take the form of what commentators call "revolving door arrangements." McGraw and Wilber, both of whom studied the Marcellus Shale gas boom in Pennsylvania, found numerous examples of oil industry staff taking up positions with state regulatory agencies and conversely state regulators accepting high-paying jobs with shale gas companies.

Zeese and Flowers documented similar examples of corporations targeting government senior officials and regulators to join the oil and gas industry; or networking and exploiting relations with industry staff who had joined government agencies. A 2011 Pennsylvania audit found that dozens of state officials had gone to work for the very industry they had previously policed, twenty-eight officials in the previous year alone. The strategy can, however, have unintended consequences. Kelly notes that Pennsylvania has gained a reputation in the United States for "its inability—and at times apparent reluctance—to police the drilling boom."[32] There may be some truth to the accusation. A recent study by Cornell University Professor Anthony Ingraffea of drilling within or on the borders of Pennsylvania's 100,000 acre Loyalsock State Forest that found that 59 percent of gas wells had never been inspected. Of the eighty-six wells that were inspected, six were leaking.

The revolving door appears to be a widespread if informal practice; the two-way benefits are implicitly understood by participants. Norway's former minister of finance, Karl Eirik Schjøtt-Pedersen, went on to become director of the Norwegian Oil and Gas Association. In the United Kingdom the *Guardian* reported that under recent Labour governments, BP gained the nickname "Blair Petroleum" because of the constant flow of personnel between the company and government positions. The practice continued under successive Conservative governments. And in one of the most famous cases, Dick Cheney was chairman of the board and CEO of petroleum industry services corporation Halliburton for five years before he became George W Bush's vice president. Prior to his time at Halliburton, he was an affiliate of conservative think tank The American Enterprise Institute for Public Policy Research.

The central purpose of corporations is to generate profits, so they have an incentive to reduce and where possible externalize their operating costs. "Externalizing" means not just avoiding but transferring as many of these costs as possible to someone else: communities, municipal authorities, central government (i.e., taxpayers), or the environment. That's where globalization becomes a useful tool. With mobile capital and differing regulatory regimes to choose from in oil-endowed countries, petroleum corporations can do business with "whatever locality offers the maximum opportunity to externalize costs through cash subsidies, tax breaks, substandard pay and working conditions, and lax environmental standards. Income is thus shifted from workers to investors, and costs are shifted from investors to the community."[33] It is ultimately about the exercise of power.

Selling the Industry to the Silent Majority

In the twenty-first century governments continue to rely on oil and gas as a key component of energy policy. At the same time, they have the difficult task of balancing economic reliance on fossil fuels with growing public concerns over environmental destruction and climate change caused by the extractive industry. The petroleum industry cannot simply rely on governments to plead its case. Companies have to devote considerable resources, time and expertise to winning over the hearts and minds of what one oil PR executive called the "Silent Super Majority."[34] This strategy is pursued through a number of tactics, including (a) exaggerating the sector's reserves and future energy role, (b) "green-washing," (c) showcasing corporate social responsibility, and (d) acknowledging problems while covering up harmful practices.

Most petroleum industry analysts expect that increasing world population (possibly another two billion by mid-century) and improved living standards will lead to higher energy demands, which will not be able to be met by alternative sources. Oil and gas will have to supplement alternative energy for years to come.[35] Those predictions are not a foregone conclusion, as we'll note shortly. To bolster their claims, many of these same people overestimate oil and gas reserves and play loose with industry technical terms to confuse the issue. This appears to be a widespread industry practice to mislead politicians and the public into supporting industry expansion so they can tap into the supposed bonanza in export earnings, royalties and jobs that lies just beneath their feet.[36] We'll see later how this particular tactic has been used by the New Zealand oil industry in cooperation with the National government to "sell" the population of the country's East Coast on oil and gas development.

Since the rise of the first environmental movements in the 1960s, global petroleum corporations have seen the value of adopting terms like "sustainability" and proclaiming their green credentials.[37] They realized that claiming to follow safe, "environmentally responsible" practices regulations was more effective than questioning the need for environmental protection. They insisted the risks were low and, borrowing a strategy from the tobacco industry, challenged evidence of environmental degradation.[38] When instances of environmental damage and criticisms continued, they shifted to supporting government claims of having put in place a rigorous regulatory regime (albeit with industry input).

Environmentalists sometimes refer to such tactics as *green-washing*. Zeese and Flowers documented how the American Petroleum Institute created multimedia public relations campaigns with innocuous names such as "Energy Tomorrow" and "Energy Citizens" to "provide a green-washed image of the industry as one that is concerned about the environment and health and that is creating jobs and energy security."[39] Karliner maintains that corporate environmentalism is in practice a two-faced strategy. In the United States it is not unusual for large corporations to "proclaim their environmental concerns and good practices, while working through their lobbyists to slash the budgets for enforcing agencies such as the Environmental Protection Agency and the Interior Department," cutting science budgets, and "undermining laws mandating clean air and water, as well as those protecting endangered species and wilderness."[40]

Corporate environmentalism is promoted and facilitated by a variety of international voluntary organizations who receive their professional aura and tangible support from being associated with large multinational corporations. A well-known example is the World Business Council for Sustainable Development (WBCSD).[41] The organization was established in the lead up to the 1992 Earth Summit to represent global business interests. Top-level executives were concerned about the potential economic and reputational damage the Summit could do to businesses if they weren't at the table to "manage" the debate.

Extractive industries have increasingly embraced the corporate social responsibility (CSR) movement that has become popular in business circles. Hilson[42] notes that multinational mining and petroleum companies operating in Africa, Latin America, and Australasia have adopted CSR as a strategy for responding to criticisms of their performance. Jenkins and Yakovleva suggest that the spread of the CSR agenda has made extractive companies more aware of the need to "justify their existence and document their performance through the disclosure of social and environmental information."[43] Nevertheless, Jenkins[44] found in examining the "sustainability" and CSR reports of several mining companies that they were long on PR spin, but short on demonstrating understanding of what made local communities tick, what gave rise to conflicts and the company's responsibility for causing and resolving them.

The CSR movement has been perpetrated by industry organizations, consultants, and even academic institutions. Queensland University's improbably named Centre for Social Responsibility in Mining within the Sustainable Minerals Institute is a case in point. The Centre serves extractive industries with education courses on sustainability and social responsibility, research reports and techniques for managing conflict and engaging with communities affected by mining. CSR is a main focus. In 2011, for example, Centre director Daniel Franks teamed up with Rachel Davis of Harvard's Kennedy School of Business to present a paper at the first International Seminar on Social Responsibility in Mining in Chile, suggesting community conflict and protest were the result of *"badly managed 'transformations' of local communities and environments,"* [!] and may lead to a complete breakdown of company's social license to operate.[45]

Owen and Kemp[46] maintain, based on a critical international review of mining and sustainable development, that the idea of a social license has been used as a survival strategy for ensuring the viability of companies and the sector as a whole. It has been of little useful benefit for communities and has little to do with sustainable development. Frynas[47] as well as Gilberthrope and

Banks[48] are similarly skeptical of adoption of CSR in extractive industries. They suggest it has been used to "legitimize the sector after decades of environmental disasters and the trampling of indigenous rights" and has produced little real development at the grassroots level.

Petroleum companies and trade associations claim to be open and transparent in their reporting and activities, but such claims often serve merely to deflect attention from risky and harmful operational practices. Corporate executives and industry advocacy groups have recognized the public relations value in speaking out for improved environmental stewardship and action on climate change, although they and their companies often act differently in private.

In the United States the petroleum industry underscores its transparency by pointing to voluntary reporting of fracking operations through the government/industry jointly funded database FracFocus. However, independent investigations by journalists and environmental organizations cast doubt on how transparent and thorough company reporting really is. One report found that although industry records show at least 85 percent of US wells are fracked, an eight-state sample of FracFocus data found that only 47 percent of wells were reported to have been fracked. Furthermore most corporations refuse to divulge all the chemicals used in their fracking operations, claiming the information is commercially sensitive and the health risks miniscule.

Engaging/Controlling Affected Communities, Municipal Councils, and Indigenous Groups

Acting like a good neighbor and promising benefits to communities has proven an effective strategy in helping petroleum companies project a positive image and defuse local protest. As Wall Street financial analyst Deborah Rogers observed sarcastically, oil and gas companies "are not in business to steward the environment, save the family farm, or pull depressed areas out of economic decline. If these things should by chance happen, there are merely peripheral to the primary mission of the companies."[49]

This strategy is pursued through a number of tactics, including (a) consultations, speaking engagements, and other trust building tactics; (b) collaborative stakeholder partnerships; and (c) buying community support.

Various methods have been used to give effect to this strategy, from invitation-only presentations to selected "stakeholders" to public seminars, forums, debates, consultations, and briefings. McGraw[50] and Wilber[51] provide first-hand accounts of how such tactics were employed by large gas corporations in the early stages of the Marcellus Shale boom. Public "information meetings" were a favorite technique for company lawyers and exploration managers, while land access managers preferred a friendly chat on a farmer's front porch. In Australia, the Centre for Sustainable Mining at Queensland University produced several research papers suggesting ways extractive companies could improve their "engagement" with mainstream and Aboriginal communities to be sure of successful outcomes for the company.

After several Colorado communities enacted bans or moratoria on fracking, Anadarko established a community relations team headed by drilling engineer Alex Hohmann. The team's brief was to attend community meetings and go door to door answering questions and listening to concerns about drilling activities, some of which were taking place literally over the back fence or in a local school yard. Hohmann stated, "We have to demystify it, to help people see it as compatible with their lives. That's our industry's challenge."[52]

North of the border where controversy over tar sands mining has gathered momentum, the Canadian Association of Petroleum Producers declared publically that its members were "committed to being good neighbors through regular consultation and communication with local landowners and affected communities." They emphasized the importance of building long-term relations with "key stakeholders, including aboriginal communities."[53]

Collaborative stakeholder partnerships (alternatively called citizens' advisory groups, community forums, or consultative committees) frequently feature in corporate community engagement strategies, usually organized in conjunction with federal or municipal governments. Most petroleum companies have experimented with such strategies at one time or another, usually linked to the company's budget for civic and charitable contributions. They reason that the alternative could be costly court action, compulsory land acquisitions, and beefed up security against protesters.

Several presenters at the 2011 Media and Stakeholder Relations Hydraulic Fracturing Initiative, a PR conference in Houston for the unconventional oil and gas industry, emphasized the importance of having a community engagement strategy including establishing local stakeholder groups. One presenter stressed that once companies had a "seat at the table," they should use it to control the dialogue or someone else would. Another speaker, possibly with a military background, observed that strategies for engaging with communities and building grassroots support were particularly necessary when drilling near "high density neighborhoods . . . *to minimize community uprising.*"[54]

There's more to such tactics than meets the eye. Karliner[55] argues they are part of a "pseudo system of corporate governance" by which residents, municipal councils, businesses, and institutions are transformed from citizens to stakeholders in a corporate enterprise. "It redefines citizens and their communities as constituencies of transnational corporations in the world economy." In effect they become involuntary shareholders, part of the corporate enterprise, presumed to share its goals and values, but with little to show for their involvement but promises that all the drilling and fracking will ultimately benefit the community.

An important tactic for large oil and gas corporations is having a carefully thought-through program of community sponsorships, prizes and donations to worthy causes, charities, sports and recreation groups, civic facilities, and conservation projects.[56] Presenters at the 2011 Houston oil and gas industry PR conference stressed the importance of having a charitable giving/outreach budget to support a company's engagement strategy. One presenter reported that his company's community engagement budget focused on planting trees, "outreach" (community funding) and staff volunteering. Apache Corporation put that tactic into action in 2012 when it leased over a hundred thousand acres in Montgomery County, Montana. Within a few months of commencing exploration, the company had made over $30,000 in charitable contributions to local community programs and voluntary groups. Targeted local community funding is helpful in building trust in a company, solidifying support, and undercutting anti-oil activists.

McGraw and Wilber both observed examples of targeted community funding during their in-depth studies of the rise of the shale gas industry in Pennsylvania. In the early days companies were often called upon to "mitigate" problems that arose from their drilling operations. Operations were on a relatively small scale, and some companies simply paid for water tests and monitoring, water delivery, and even installed expensive filtration systems when landowners complained of contaminated wells. Easier and less expensive than costly litigation.

In Latin America, Widener[57] found that "the reaction of the oil industry and the state to community and NGO efforts varied." Oil multinationals responded to protest action by communities affected by drilling by providing community services and support for conservation projects, as well as taking greater operational care near ecologically sensitive areas that were under public scrutiny. "The state responded both with force [but also] with community projects for the oil-impacted communities, while deriding the environmental concerns of the environmental activists."

Elsewhere, corporate "community investment" tactics have been less subtle. In a number of instances in the United States, companies have combined legal threats with offers to buy out obstructive landowners. Sometimes the deal-making takes place with the local authority. In 2014 pipeline company TransCanada Corporation gave the Ontario town of Mattawa C$30,000 under the company's community engagement program. The only proviso was that the municipal council would agree to "not publicly comment on TransCanada's operations or business projects" for five years.[58] That conveniently covered the period during which the company was planning to build a cross-Canada oil pipeline called Energy East that was planned to pass near Mattawa. Critics labeled the proviso a paid gag order.

Elsewhere McNab and her colleagues[59] documented recent changes in "corporate social investment" behavior in the Australian extractive resources sector. Traditionally such donations, program grants, and civic contributions had at least a quasi-philanthropic purpose. Lately, the researchers noted, they had become more agenda-driven and defensive in an apparent attempt to mitigate "social risk" to the company from increased environmental lobbying and protest action.

Defending Company and Industry Practices

Fossil fuel companies on a global scale are investing more time and money trying to convince governments and publics that they are safe, environmentally responsible, and essential to a country's long-term economic development, possibly because they fear being labeled a pariah industry and regulated out of existence. Their less scrupulous promotional and defensive tactics have certainly come under closer scrutiny in recent years. This strategy is pursued through a number of tactics, including (a) manufacturing debate and controlling public discourse; (b) harnessing mainstream media; (c) utilizing disinformation, "independent experts," and academic institutions; (d) bankrolling conservative foundations, think tanks, and advocacy groups; (e) exploiting social media; and (f) infiltrating the education system.

A number of studies have documented how the petroleum industry has borrowed from the playbook of the tobacco industry, attempting to stimulate debate about the importance of oil and gas development and sewing doubt about the harm the industry does. Oreskes and Conway[60] describe how from the 1960s and 1970s, Big Tobacco "manufactured debate" over scientific research about whether smoking caused cancer, convincing the mass media that responsible journalists had an obligation to present a "balanced" argument. Big Tobacco engaged supposedly independent scientists (with funding channelled through lawyers and foundations) who were prepared to design studies in such a way that their results favored the industry. Over time they constituted a body of "expert" witnesses who could be called upon to speak in support of industry or at least present an alternative view of the evidence.[61] The public had no way of knowing that the "evidence" being presented in a supposedly balanced public discussion was "part of an industry campaign designed to confuse."[62] Such campaigns convinced people who didn't know otherwise that there was still considerable doubt over the matter of whether smoking caused cancer. "The industry had realized that you could create the impression of controversy simply by asking questions, even if you knew the answers and they didn't help your case."[63]

Oreskes and Conway point out that such tactics have been picked up by the oil and gas industry to address increasing public concerns over health effects, environmental damage and climate change. Dunlop and McCright came to a similar conclusion: "The conservative movement/fossil fuels complex quickly adopted the strategy of 'manufacturing' uncertainty and doubt . . . as its preferred strategy for promoting skepticism regarding climate change."[64] The Union of Concerned Scientists[65] called the tactic "misrepresentative information."

American PR guru Richard Berman, in a speech to the Western Energy Alliance Conference in June 2014,[66] explained the technique of reframing the debate and fostering skepticism:

> I tell you, when I'm on offense, I'm going to reframe the issue. I'm not going to allow the conversation to be based on how somebody *else* has framed the issue. Because then I'm on the defense. I'll be arguing over what *they* said. Repositioning the opposition suggests telling people, 'Oh you think that this group is a group that does X? Well, let me tell you, what they are *really* doing is Y. I don't care what they tell you they are doing, they're doing something else!' Challenge the legitimacy of local protesters and environmental activists to speak for the wider public.

A prime example was provided at the 2011 Houston oil and gas industry PR conference. A regulatory affairs manager for a large international petroleum corporation argued that in-house PR managers were the communication brokers between company technicians, regulators, and the public. They shouldn't leave it to the media. They needed to "repackage" scientific and operational information (e.g., about fracking) so it was understandable to the lay person, enabling the company to get its key messages across. It was basically a sales job. For example, to answer concerns about fracking fluids, this executive said:

- "We talk about biocides. Wow, that's a big word. But that's bleach. So we've got to start talking bleach. . . .
- "Tell them polyacrylamides are ChapStick. . . . People get really worried when you talk about putting polyacrylamide in frack fluids that they're gonna pump down a well. What is that? It's ChapStick—so take out your stick of ChapStick when somebody talks about polyacrylamides.
- "Tell them surfactants are Dove Soap!"[67]

Richardson and Denniss define such information repackaging as "the art of ensuring people view a situation from the 'right' perspective."[68] Controlling the debate enables the oil and gas industry to shift the focus from *corporate interest* to *national interest*.

The oil and gas industry maintains that media reportage is influenced by misinformation and fear-mongering. To address this situation, corporations have cooperated on their own environmental campaigns, engaged independent-appearing experts, and aggressively disseminated selected "factual" information to reporters, editors, and publishers. Oreskes and Conway[69] argue that, until recently when investigative journalists began to do some serious digging, such campaigns were accepted with little critical examination by a largely acquiescent media.

One way the oil and gas industry learns and shares state of the art techniques for manipulating mainstream media is through industry symposia and conferences like the 2011 Houston PR conference, which was targeted at unconventional oil and gas companies. The first day's session was about "*Harnessing* mass media, social media and grass-roots community support to develop dynamic strategies for overcoming public concern over unconventional oil & gas production." Several PR executives shared tips on dealing with the media in a session on "Media Relations Strategy":

- Presenter D urged his listeners to "inoculate reporters early" against what he termed "a strong activist insurgency" occurring across the country, presenting the company's message through "factual" releases, briefings, and visits to showcase drilling sites laid out like movie sets. He recounted how a reporter for *Forbes* magazine received his "inoculation" by being flown to company headquarters to rub elbows with top management, and then was whisked away on a helicopter trip to a showcase well pad equipped with "best practice" technology. The aim was to leave the reporter with the overwhelming impression that the company cared (about people and the environment) and "we know what we're doing."
- Presenter E discussed techniques for engaging and educating a "misinformed media" by being more forthcoming and other tactics to "proactively combat negative press."
- Presenter F enjoined his listeners to "push out one pro-active item each week." Company PR staff could use the media to control the public debate by writing essays, commentaries, and letters to the editor, and answering every critical letter to the editor.[70]

Korten, Heinberg, Van Gelder,[71] and others point out that mainstream media are under pressure to go along with such strategies since they are beholden to wealthy corporate interests through their advertising budgets and their boards. It's a difficult balancing act because their public credibility is on the line. Chomsky and Herman[72] concur that structural factors shape the role of the media: "Ownership and control, dependence on other major funding sources (notably advertisers) and mutual interests and relationships between the media and those who make the news and have the power to define it and explain what it means."

Far from ducking controversy over climate change and the negative impacts of oil and gas production, the petroleum industry is confident that it has the experience, financial resources, and hired expertise to be able to discredit conventional science, cherry-pick facts for disinformation campaigns, and manipulate public discourse. This is particularly crucial in regard to debates over fracking and climate change. According to Dunlop and McCright, the conservative movement in the United States and several other countries has joined forces with the fossil fuel industry to attack climate change science because it could lead to carbon taxes and government regulations that could cripple the industry. Since the release of a series of reports by the Intergovernmental Panel on Climate Change (IPCC), their tactics have extended "well beyond manufacturing uncertainty, increasingly criticizing peer-reviewed research reports, refereed journals, governmental grant making, scientific institutions (e.g., the US Academy of Sciences) and the expertise and ethics of scientists."[73]

Examples of dubious research and shady funding arrangements with supposedly reputable scientists and academic institutions abound. Zeese and Flowers[74] cite the example of the Shale Resources Sustainability Institute which was established at the State University of New York in Buffalo. "Fortunately, their ties to the industry were discovered and when media exposure after their first study was released demonstrated clear bias, the program was closed by the university."

Indeed, the number of industry-linked academic centers has increased dramatically in recent years. Such centers and industry-contracted research are often based in university science faculties or business schools. Some of these appear legitimate, but their aims and research programs are skewed toward corporate interests. They may simply fail to acknowledge the indirect support they receive from extractive industries. For instance, the University of Pennsylvania's Marcellus Center for Outreach and Research was established to undertake science-based research to aid understanding of the issues surrounding the shale energy development. Stakeholders include government, industry, communities, and environmental groups. The Center is internally funded within the university, although the university as a whole receives substantial endowments from the oil and gas industry. According to the Center's

Director, Dr Tom Murphy, [75] their aim is to inform and educate stakeholders about the benefits and risks of shale gas development. Murphy distinguishes between "true risks" and the "perceived risks" communities often worry about. MCOR's program is based on four premises: "(1) Shale gas development should and will go ahead. The benefits outweigh the 'risks.' (2) Shale gas production is safe and well-regulated. (3) Communities and environmental groups are often ill-informed and need to be educated. (4) Climate change is not our problem."

In a similar vein, the University of Queensland's Centre for Social Responsibility in Mining within the Sustainable Mining Institute is "committed to improving the social performance of the resources industry globally." The Centre claims to have "a long track record of working to understand and apply the principles of sustainable development within the global resources industry." An introduction to the Centre's Managing Cumulative Impacts project states:

> This project aims to enhance *mining industry efforts* to proactively address the management of cumulative environmental and socio-economic impacts at the community and regional scale and *strengthen the capacity of the industry* to engage with planners, regulators and others in multi-stakeholder processes for monitoring and managing such impacts associated with mining. [76]

The fossil fuel industry has been sponsoring and utilizing university research in the United States and other countries for decades. Richard Shiffman [77] investigated several industry-linked research programs and concluded there were serious questions about whether the studies could be trusted. One of the most celebrated cases involved renowned academic and climate change denier Dr Wei-Hock Soon of the Harvard-Smithsonian Center for Astrophysics. Soon was accused of accepting money from oil companies and the American Petroleum Institute to produce peer-reviewed articles on climate change for scientific journals without acknowledging his sponsorships. [78]

Establishing and financing sympathetic foundations, think tanks, and advocacy groups is another widely used petroleum industry tactic to influence public and political thinking around fracking, climate change, and clean energy. Dunlop and McCright observed that conservative foundations and think tanks typically share a neoliberal ideology: "a universal commitment to free enterprise, limited government, and the promotion of unfettered economic growth." [79] In the United States, some of these conservative institutions have supported research and reports critical of mainstream climate change science because of the regulatory implications for business. They have served as a convenient screen for corporate interests who wish to anonymously support climate deniers and deflect calls for rapid transition to a low-carbon economy.

An investigation by Frumhoff and Oreskes[80] found that some of the largest corporations such as Shell, Chevron, and ExxonMobil publically accepted climate change science while "continuing to support climate denial through [funding] influential lobbying groups and trade associations."

The petroleum industry has embraced social media as an important tool for circulating positive spin about the petroleum industry and subtly manipulating public debate. The literature suggests at least some of the larger petroleum corporations are contracting IT specialists, tapping into the blogosphere, collaborating with right-wing bloggers, and encouraging the active involvement of their company employees and acquaintances to get positive messages out and answer critics.

Several presentations at the 2011 Houston PR conference showcased ways different corporations were utilizing company websites, staff blogs, Facebook, Twitter, and YouTube to spin positive messages and build up an online clientele. A whole session at the conference was devoted to "Leveraging Social Media as a Communication Tool for Educating and Engaging Stakeholders to Counter Myths Surrounding Unconventional Oil & Gas Production." For example:

- Presenter G stated that his company had a social media team comprised of three full-time social media staff, seven part-time staff, and various regional social media "leaders" (paid bloggers and pro-industry commentators). For expert advice and information, they also had departmental consultants in legal, regulatory, government relations, operations, and field services as well as paid PR consultants. The specialists analyzed people who engaged with the company's various social media platforms, how they responded, and what their interests were so the company could better tailor its messages. He also recommended "marshalling" a company's staff, relatives and friends for its social media campaigns "since they have the most to gain."
- Presenter H spoke about how her company used social media to engage stakeholders and "drive public education."
- Presenter I, a corporate communications consultant, highlighted her company's social media activities and recommended four key tactics: (1) humanize the industry, (2) connect with your stakeholders, (3) lessen reliance on the (mainstream) media, and (4) localize your message. In a subsequent session, she described online tools her company used to monitor public perceptions of the petroleum industry and "quell unfounded criticism." [81]

Korten[82] notes that corporations "are moving aggressively to colonize the second major institution of cultural reproduction, the schools." They utilize pseudo-educational campaigns and produce "educational" materials to promote their industry to parents and the general public as well as young people. Faced with an environmentally conscious public and calls for greater industry accountability and regulation, Karliner observes corporations have begun "infiltrating educational systems in an effort to pre-empt the emergence of a new generation of critical environmental activists."[83] The strategy, which dates back to the seventies in the United States, is now spreading to other countries. It includes providing scholarships and sponsorships, endowing academic chairs, and funding new buildings and equipment.

The tactics are becoming increasingly more sophisticated, particularly at primary and secondary school levels. Karliner recounts how after a court awarded 20,000 plaintiffs $5 billion in damages following the Exxon *Valdez* oil spill, Exxon distributed a free video to 10,000 primary schools denying there were any lasting effects on the local ecosystem or people's livelihoods. In another instance an investigation by the *Philadelphia Enquirer* found that school lesson plans about the environment typically included materials supplied (free) by corporations like Proctor and Gamble, Chevron, and ExxonMobil with acknowledgments to the companies. Chevron, for example, distributed a box set of interactive activities so students could learn about the industry and make "real-world energy choices."[84] Government constraints on education funding in some countries have led schools to more readily accept corporate-produced curricula.

Combatting Opponents

While the oil and gas industry is busy trying to win the hearts and minds of the silent majority, they also have to contend with the growing influence of local anti-petroleum groups and environmental movements concerned about issues like fracking, environmental disasters, and climate change.[85] The activities of these groups have the potential to damage the industry's public image, invalidate their "social license to operate," and make exploration and production more costly. This strategy is pursued through a number of tactics, including (a) countering protest groups; (b) coopting moderate environmentalists and marginalizing "extremists"; and (c) utilizing the judicial system, legislation sanctions and police action.

The industry has adopted a variety of tactics to learn about, subvert, and defeat local protest groups. One of the presenters at the 2011 Houston petroleum PR conference claimed a "paradigm shift" was occurring in the industry, from being defensive and uncompromising in dealing with opponents to listening, understanding and being more transparent. Another speaker maintained that it was basically a matter of countering the emotional arguments of oil and gas opponents with safe industry practice and scientific facts.

At the community level, Duncan and McCright found that corporations, trade associations, and front foundations have adopted the tactic of establishing and funding "astroturf organizations" disguised to look like authentic grassroots groups. "They are created to lobby or campaign on behalf of their sponsors, who hope to remain hidden from view. . . . Astroturf efforts come and go in response to specific events and policies."[86] In 2009, the American Petroleum Institute created and funded "Energy Citizens." Since then, the organization has grown into a network of local chapters all across North America. One of its recent initiatives was to organize "citizens' rallies" against climate change legislation in more than twenty states. The Energy Citizens program has been picked up by the Canadian Association of Energy Producers.

An investigation by the *Guardian* in late 2014 revealed that the Western States Petroleum Association (WSPA) was carrying out a stealth campaign to block climate policies in California. One of their main tactics was backing a network of astroturf groups with names like "California Drivers Alliance" and "Californians Against Higher Taxes."[87] Shell, BP, Chevron, and Exxon-Mobil are all members and funders of WSPA.

Sometimes the tactics are more clandestine and ethically dubious. Matt Charmichael, External Relations Manager at Anadarko Petroleum, began his presentation to the 2011 Houston petroleum industry PR conference by recommending that delegates "download the U.S. Army/Marine Corps Counterinsurgency Manual, because *we are dealing with an insurgency.*"[88] He was preceded by Range Resources communications director Matt Pitzarella, who spoke about overcoming stakeholder concerns around fracking:

> We have several former psy ops [psychological operations] folks that work for us at Range because they're very comfortable in dealing with localized issues and local governments. Really all they do is spend most of their time helping folks develop local ordinances and things like that. But very much having that understanding of psy ops in the Army and in the Middle East has applied very helpfully here for us in Pennsylvania. These people are important in attending hearings and maintaining public trust amidst special interests *that often use misinformation to create fear.*

Others in the public relations industry have adopted a similar military campaign approach. PR guru Rick Berman was quoted as telling a Western Energy Alliance conference in Colorado Springs to "think of this as endless war."[89] Company executives "must be willing to exploit emotions like fear, greed and anger and turn them against the environmental groups. And major corporations secretly financing such a campaign should not worry about offending the general public because 'you can either win ugly or lose pretty.'"

Petroleum companies and their PR consultants also undertake covert research and surveillance of environmental organizations, anti-fracking movements, and individual activists in order to gather incriminating information and learn their next moves. Jack Hubbard, a Berman and Company vice president, described to the Western Energy Alliance conference how he had carried out detailed research on the personal histories of members of the boards of the Sierra Club and the Natural Resources Defense Council to try to find information that could be used to embarrass them. Similarly, Wilber[90] recounts the story of a group of rural, middle-aged Pennsylvania women who became activists against Marcellus Shale gas companies when they became concerned about the health and environmental impacts on their community. They discovered they were under surveillance by police agencies after they were labelled "anti-drilling environmentalist militants" and placed on an Office of Homeland Security watch-list. According to an investigation by the Earth Island Institute, the US gas and oil industry not only has its own security apparatus but enjoys close relations with law enforcement and federal security agencies in information sharing, planning and operations against environmental extremists.[91] Companies operating in more volatile areas of the world place a high priority on such arrangements.

Since the 1992 Earth Summit in Brazil, extractive industries have experimented with a two-prong strategy of courting cooperative relations with middle-of-the-road environmental organizations while attempting to discredit and marginalize "radical" organizations. At the 2011 Houston petroleum industry PR conference, the director of an industry-funded advocacy organization[92] suggested a number of techniques for identifying the strategies being used by environmental NGOs and developing "proactive counter-strategies to combat their influence." Illustrating how to divide moderate environmentalists from extremists, PR consultant Jack Hubbard told the 2014 Colorado Springs Western Energy Alliance conference that his company had just launched a campaign in Pennsylvania and Colorado against fracking opponents which they called "Big Green Radicals." "There is nothing the public likes more than tearing down celebrities and playing up the hypocrisy angle," Mr. Hubbard said, citing billboard advertisements featuring environmental activist Robert Redford. The billboard caption read: "Demands green living. Flies on private jets."[93] On the other hand, Hubbard maintained "responsible" environmental organizations could be worked with as long as they were prepared to play a "constructive role."

For instance, after a plethora of municipal fracking bans and court cases in Colorado, oil and gas companies decided to try to be more transparent and engage with communities. Taking point in this tactical shift was the Colorado Oil and Gas Association (COGA) and its CEO Tisha Schuller.

Schuller, a geology graduate with fifteen years' experience arranging permits for oil and natural gas projects, claimed to be a converted "environmental activist."[94] Schuller toured Colorado communities delivering what she called "Deescalating the Fracking Wars" talks.[95] She maintained that the debate over drilling has become polarized between hardline industry proponents and anti-fracking protesters. Each tried to score points over the other to win adherents. For community activists in the Saul Alinsky tradition,[96] confrontation is a legitimate response to an exploitive, oppressive force. For Schuller (acting for "the oppressor"), the trick to winning the hearts and minds of communities was to render polarized conflict and confrontation illegitimate. In other words, subvert the power of confrontational, social action-oriented groups by refusing to engage with them. Schuller and COGA staff did this by presenting themselves as caring citizens, neutral moderators of debate, and purveyors of trustworthy factual information. Schuller said, "All the interesting work happens when opponents meet in the middle."[97] In fact Schuller's tactics ensured there were no longer any "opponents," just concerned citizens. If you are in the ring, you've agreed to play by COGA's rules. If you're obstinately anti-fossil fuel, you're simply not in Schuller's arena.

To accomplish this subversion:

- Schuller and COGA staff visited over thirty communities in three years to get to know various stakeholders and councils and their concerns.[98]
- COGA staff went out of their way to make sure protesters on the street had sunscreen, water and energy bars. "We try to be kind—we don't want to energize a protest and make it a story."
- Since professional activists are often able to mobilize quickly via social media, one of Schuller's staff constantly monitored activist websites. When a protest is planned, COGA alerted a network of friendly business groups and trade organizations not directly connected to oil and gas, and asked them to speak out in support of the industry.
- COGA wasted little time on "extreme activists who threaten industry workers or damage property" since they can't be "talked around" to the industry's point of view.

Schuller said they preferred to let the media portray the most militant activists as the face of protest. "We're wanting the 80 percent in the middle,"[99] the Silent Majority.

Karliner observes that corporations have been "highly successful in drawing environmental groups into dialogue."[100] While some of these encounters have produced working accords and operational changes, they have frequently come "at the expense of affected local communities and grassroots organizations"[101] leaving them feeling that they've been sold out. Hard-line environmental activists, on the other hand, tend to view such cozy relations with big business polluters as at best co-optation and at worst collusion.

Where there is known to be strong community opposition to a company's exploration and production activities, companies usually liaise with government and local authorities on law enforcement measures. Protracted protests can impose significant costs on companies, and in some countries the state police and military apparatus can be mobilized to ensure operations proceed. In open democracies private security contractors are more often the preferred line of defense even though they must operate under stringent legal constraints. When a persisting threat of protest action requires additional tactics, companies can resort to other kinds of state-sanctioned force. Setting aside covert (and often illegal) operations, in most countries this usually means the threat of emergency legislation, legal proceedings and/or police enforcement.

McGraw and Wilber document instances during the early phase of the Marcellus Shale boom in which gas company lawyers threatened legal action against protest groups or hold-out landowners to force them to comply with company plans. But court action and legislative sanctions are also being used against local authorities across North America who impose moratoria or bans on unconventional drilling and fracking within their boundaries. More than four hundred communities from California to New York have passed measures against fracking, according to Food and Water Watch. In response, the petroleum industry—often in cooperation with state governments—has moved to curtail community initiatives against intrusive oil and gas operations and fracking. As the result of court action and legislation, local communities in many US states no longer have the right to ban fracking and have little power to regulate unconventional oil and gas operations within their municipal boundaries.[102]

For instance, in Colorado four municipalities—Fort Collins, Boulder, Lafayette, and Bloomfield—passed referenda banning fracking in 2013. The City of Longmont passed a fracking moratorium shortly afterward which also banned the storage and disposal of fracking waste within city limits. COGA, the industry trade group headed by Tisha Schuller, spent over a million dollars in media campaigns and other activities to try to defeat the referenda. Shortly after the bans and moratoria were announced, COGA and a state agency initiated joint court action that eventually saw them overturned.

Meanwhile in Texas, the petroleum industry tried to slow a similar wave of local community bans and regulations. Residents of Denton, Texas, for example, were concerned about the growing number of unconventional wells and fracking within city limits causing noise, traffic, air pollution, and health risks. In spite of an expensive industry-funded campaign and death threats, a group of resident activists managed to get an anti-fracking referendum passed in November 2014. Within weeks, two Texas state agencies joined a petroleum industry lawsuit to overturn it. A group of conservative politicians introduced a bill into the Texas state legislature (HB 40) in March 2015 aimed at overriding Denton's ban and similar bans by municipalities across the state. The bill passed the Texas state legislature in July 2015.

SOCIAL CONFLICT AND CHANGING INSTITUTIONAL CONFIGURATIONS: THE CASE OF OIL AND GAS

Benson and Kirsch[103] introduce the notion of "the politics of resignation" to explain how corporate defense strategies contribute to citizen apathy, mistrust, and a lack of civic engagement. But it is also useful, as Bebbington's recent volume on South American extractive industry suggests, to explore the structural effects (institutional political and economic changes) that can result when corporations employ strategies to solidify government relations, deflect public criticism, and combat environmental activism and community protest movements.

Since the 1970s two streams of social/economic thought on managing production and providing for societal wellbeing have contended. One, called Neoliberalism (e.g., Thatcherism, Reaganomics), emphasizes the central role of the free-market and the need to reduce regulatory and bureaucratic impediments to the market's efficient functioning. Several other strands of thought adopted an alternative perspective collectively known as Corporatist theories. They emphasize the essential role that institutions play as coordinating and control mechanisms regarding the market. These institutional relations evolve as a distinctive *social structure of production, distribution, and accumulation* in the interplay between the state, the private sector, and civil society. Godelier emphasizes that such institutional linkages occur through certain agreed rules (laws).[104] In effect each particular institutional configuration is the result of a kind of social pact between powerful groups with different interests who find it to their benefit to cooperate in responding to changing economic and social circumstances by developing policies, establishing rules and regulations, and agreeing priorities and expenditures. Changes in the social system of production (i.e., the configuration of institutions) that evolve over time in different nation-states account in large part for differences in their development trajectories. Another name for this institution-focused approach is Regulation theory.[105]

Regulationists Hollingsworth, Rogers, and Boyer[106] give as examples of institutions the labor relations system, the training system, the internal organization of firms, industry associations and partnerships, structured relationships with suppliers and customers, financial markets, courts and law enforcement, and the organization of the government and its policy-making. Borrowing from anthropology, what gives a social system of production and accumulation its distinctiveness in any given country is the fact institutions are embedded within and influenced by a society's shared symbols, customs, and traditions, as well as behavioral norms, moral principles (equity, fair play), rules, and laws. Even in more diverse societies there is usually some form of consensus, albeit constantly renegotiated, about the values, principles and norms that should guide how these economic, political, judicial and social institutions relate and their ultimate aims. In New Zealand such values include the Kiwi sense of justice and fair play, continuing to honor the Treaty of Waitangi made between European settlers and Māori, and pride in "our clean, green image." As Hollingsworth, Rogers and Boyer explain:

> All these institutions, organizations and social values tend to cohere with each other, although they vary in the degree they are tightly coupled with each other into a full-fledged system. While each of these components has some autonomy and may have goals that are contradictory to the goals of other institutions with which it is integrated, an 'institutional logic' in each society leads institutions to coalesce into a complex social configuration. This occurs because institutions are embedded in a culture in which their logics are symbolically grounded, organizationally structured, technically and materially constrained, and politically defended. The institutional configuration usually exhibits some degree of adaptability to new challenges, but continues to evolve within an existing style. But under new circumstances or unprecedented disturbances [e.g., major protests or civil unrest], these institutional configurations might be exposed to sharp historical limits as to what they may do and not do.

From a regulatory perspective we can conceive of the social system of production, distribution, and accumulation as a triangle of contending or collaborative relationships between the private, public, and civic sectors (and powerful interest groups within them) which change over time and shape the nature of institutions, the norms and values that guide institutional operations and interactions, and the policies and regulatory regimes that are put in place. The specific character of this institutional configuration and the outcome of its activities depends on which sector or powerful combination of groups (e.g., the oil industry) are in a dominant position at any given time, the prevailing circumstances and the extent to which the interests of these groups happen to coincide or diverge from other sectors (see figure 1.2).

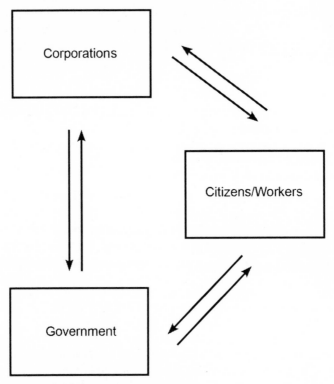

Figure 1.2.   The Regulation Triangle.
Source: Thomas J. Rudel, Timmons Roberts, and JoAnn Carmin, "Political
Economy of the Environment," *Annual Review of Sociology* 37 (2011): 228.
Reproduced with permission of Annual Review of Sociology, Copyright ©
2011, *Annual Reviews.*

For example, a study by Fisher[107] of national differences in regulating
greenhouse gas emissions found that variations in regulatory corporatist or-
ders explained most of the differences in climate policy between countries.
In Japan the government played a dominant role, in the Netherlands environmental NGOs had considerable influence, while in the US large corporations
occupied the most powerful position. The differences in institutional configu-
rations between these countries are reflected in different regulation diagrams.
In Japan, for example, the arrows from government to corporations and
citizens are large and bold to indicate dominance of institutional relations.

This brings us to the question of the relationship between governments and
the petroleum industry in countries which happen to have large oil and gas reserves. In many instances this abundance has led to what economic development literature calls "the resource curse," a term coined to indicate the macro-
economic problems and state regulatory issues that can arise when a country

becomes dependent on resource-based exports. From the standpoint of the national economy, an oil and gas boom can lead to a rise in the real exchange rate, thereby squeezing the profitability of other tradable-goods and services sectors resulting in a lopsided economy in which industrial production for the home market tends to atrophy.[108] It can also harm other sectors by competing for labor, monopolizing specialist services and suppliers, and creating spill-over effects that damage the natural resources upon which other industries rely (land, water, or scenic beauty). Bertram points out that a minerals boom can also undermine the rigor and transparency of a country's regulatory and policy-making system, as well as the integrity of the government. This is because of the economic importance of petroleum companies to the national economy and government's coffers, and their ability to fund expensive lob-bying for special favors (tax breaks, favorable legislation), a process known as "rent seeking."[109] The resulting institutional configuration, in terms of our regulatory diagram, would show the relational arrows weighted to reflect the economic and social dominance of Big Oil as in figure 1.3:

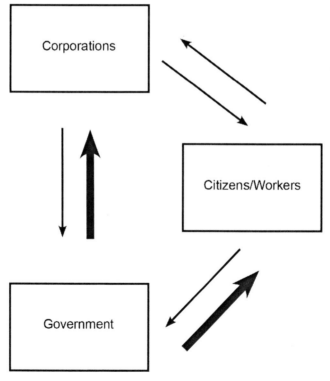

Figure 1.3. Regulatory Configuration in Countries Suffering the "Resource Curse."

Such institutional patterns are fairly typical in countries where extractive industries dominate the economy and local politics. Bebbington noted that several South American regimes which had experienced protracted mining and petroleum booms showed a clear convergence of policies and strategies in regard to resource extraction. The main elements of these strategies were "legislative changes to facilitate the expansion of extractive industry; tax and ownership changes . . . to increase the share of revenue coming to the state; use of these rents from extraction for social and other investment; and *political and discursive practices that suggest an increasingly intolerant attitude to protest and debate.*"[110] These governments found themselves forced into at best collaborative and at worst coercive relationships with extractive industries, and having to bear the responsibility for managing community relations and controlling local protest movements that could disrupt exploration and petroleum production

The resource curse may not be a serious issue for New Zealand, at least not yet, but Bebbington's summary of government strategies among states heavily influenced by extractive industry does bear striking similarity to maneuvers recently adopted by New Zealand's National government. No doubt with such trends in mind, Bertram warns that given the experience of Think Big projects to be discussed later, "policy-makers need to tread carefully and the wider public needs to insure that the nation's policy making and regulatory institutions are not captured and distorted by industry rent-seeking."

Indeed it is typical for "frontier" countries in which the discovery of mineral riches creates a boom to experience a dramatic transformation in their regulatory institutional configuration over a short period from an open democracy with a relatively balanced economy to one dominated by extractive industry. McBeath and his colleagues studied the history of Alaska's oil boom which began almost fifty years ago. In a few decades Alaska went from a relatively modest scale, mixed economy (gold mining had waned) run by the state subsidized by US federal assistance to an economy dominated by a Big Oil with all the trappings of wealth, power and influence over state government legislation, tax regime, and regulatory practices. Communities and indigenous groups received direct handouts in the form of annual per capita payments from oil royalties, but were denied involvement in big policy issues regarding development and the environment. For a while people paid little attention to the activities of the petroleum industry. Recently however there has been an upsurge in community and indigenous activism, supported by environmental allies from outside the state, concerned over destruction of the wilderness and widening economic disparities. This social conflict is putting pressure on the state government and beginning to reshape institutional arrangements, values, and priorities between the political establishment and the petroleum industry.

In cases like Alaska, similar to some of the South American studies in Bebbington's volume, we are now talking about how grassroots action (e.g., environmental movements and local struggles over oil and gas development) can alter relations between communities and extractive corporations, influence government policy-making, and alter the broader configuration of institutional relationships that controls and manages a country's production and accumulation. Ross found that resource-rich governments tend (at least initially) to defend their position and that of extractive industries when challenged by environmentalists and community protest groups. [111] They do so by becoming more authoritarian, using low tax rates and patronage to dampen demands for greater democracy, strengthening legislation, and tightening security to repress troublesome NGOs and popular movements. Even moderate regimes looking to attract and promote petroleum development (like New Zealand, as we'll see) can become less consultative, by co-opting, marginalizing or subverting civil sector NGOs, environmental groups, and community organizations.

Indeed Kirsch reminds us, "for the state to learn from social movements [and countenance institutional change] it must be willing to acknowledge the legitimacy of their concerns, or at the very least be willing to establish some kind of pragmatic détente."[112] This is more likely in countries where people's livelihoods are less dependent on a single industry such as oil and gas, and/or where there is political leeway for public debate and expressions of popular unrest over worsening social ills or environmental degradation. Even where this is the case, Rudel, Roberts, and Carmin observe that "states have typically created corporatist [interest group] policymaking circles that include long-established, moderate environmental nongovernmental organizations and exclude disadvantaged and unorganized peoples."[113] But it is equally likely for states strongly influenced by dominant corporate interests like agriculture or Big Oil to use such institutional maneuvers to co-opt mainstream environmental organizations (who treasure a seat at the table and/or are dependent on government funding) and promote themselves to the general public as environmentally responsible.

However, a problem arises for petroleum corporations and governments when local residents and marginalized groups (e.g., indigenous peoples) whose well-being, livelihoods, and culture are directly affected by drilling and production operations find their concerns are not being addressed by such high-level consultative arrangements, take matters into their own hands, and choose to remain apart from prevailing institutional arrangements between the state, corporations, communities, and NGOs. Being outside conventional structures and principles of how opposition to government policies and corporate practices should be "properly" expressed, they are in a potentially powerful position. Particularly when they are not being listened to and feel they have no alternatives, receive encouragement and support from other communities and international allies, and decide they must take action themselves.[114]

The other threat posed to established institutional relations between states and petroleum corporations arises from a growing body of unorganized but environmentally concerned consumers particularly in developed countries who take "preserving the bush," environmentalism, and climate change seriously for their own sakes and their children. Rudel, Roberts, and Carmin maintain "their power to reshape capitalism is potentially massive."[115]

CONCLUSION

In this chapter we noted that the growing power and influence of the petroleum sector is due not only to the immense wealth of large conglomerates and the world's dependence on fossil fuels, but to the process of globalization. In less developed countries with significant oil and gas reserves but lacking strong democratic traditions, this has often led to what commentators have termed the "resource curse": an economy dominated by transnational oil companies with weak governance institutions, inordinate rent-seeking behavior by petroleum corporations, social inequality, and a marginalized civil sector. But the relative strength of the global petroleum industry, the lack of democratic traditions in certain oil rich countries and the dynamics of the global energy market don't entirely account for why and how governments collaborate with corporations to promote oil and gas development or the different development trajectories that countries with significant oil and gas resources have experienced.

One factor is the particular way in which oil and gas corporations choose to exercise their power and influence in a given country, by means of a "portfolio" of strategies and tactics, taking account of prevailing governance arrangements, economic conditions, and social circumstances. We identified five broad strategies and associated tactics commonly employed throughout the petroleum industry, noting that several may be pursued at once and they can be quickly altered as circumstances change.

The second factor is the transformational, or more commonly evolutionary, structural changes that can result from the interaction between governments, petroleum corporations, and grassroots opposition movements. From the perspective of Regulation theory, we noted that the institutional relations within a given jurisdiction evolve as a distinctive social structure of production and accumulation in the interaction between the state, the private sector, and civil society. A country's particular institutional configuration is the result of a social pact of sorts between powerful groups with different interests who find it to their benefit to collaborate in responding to changing economic and social circumstances by developing policies, establishing rules and regulations, and agreeing priorities and expenditures. Changes over time in the social system of production, distribution and accumulation that nation-states evolve in part accounts for differences in their development trajectories.

Institutional relations, and the values, principles, and priorities that guide them, are not always determined by the state of relations between the governments and corporations, or the powerful interest groups behind them. We noted several instances in which social conflicts over oil and gas development, the promotional and defensive strategies employed by corporations, and the emergent power of environmental movements and local protest groups have caused governments to rethink their relations not only with corporations but with the community sector. On the one hand, studies by Ross and Bebbington noted a tendency for resource-rich governments committed to extractive-based development to become more authoritative and intolerant of protest and debate, and use mining and drilling revenues to defend the industry, win popular support and curtail social unrest. This may be a particular, perhaps later, phase in state strategizing and institutional transition especially if there has been a reasonably strong democratic tradition prior to an extractive boom. Rudel, Roberts, and Carmin observed that under certain circumstances, environmental conflict and anti-oil protest led states to establish more consultative (corporatist) policymaking circles that included moderate environmental NGOs while excluding more radical groups and disadvantaged peoples.

New Zealand is still a "frontier" oil and gas region, particularly for unconventional exploration and production, and it is early days yet in the National-led government's campaign to expand the industry. Nevertheless it is clear that for a relatively small economy, globalization processes (symbolized most recently in the Government's ratification of the TPPA) have added impetus to the National government's adoption of a Business Growth Agenda in order to expand export-led growth in partnership with multinational corporations. To attract international investment and facilitate extractive sector expansion, the Government has undertaken moves which are orthodox from the perspective of neoliberal development policy, but which have had the flow-on effect of reducing the ability of local communities and regions to manage their own planning and development. There are signs, particularly in regard to the battle to reform the Resource Management Act (RMA) that major transformations are occurring in how key institutions dealing with economic development and natural resource management relate to one another. We will explore the nature of these changes in more detail in later chapters.

We will also examine the kinds of strategies the New Zealand oil and gas industry has adopted to strengthen its relationship with the state, build a positive image in the public's eye, and deal with local protest groups and national environmental campaigns. Many of these are copied from overseas, but some of the tactics employed are home-grown. Given the petroleum industry's relatively low national profile, minimal impact on communities and Māori rohe (customary territories) outside of Taranaki to date, and the fact there has been a persistent though not militant nation-wide environmental movement opposing petroleum expansion, we might expect the industry as a whole to be employing strategies consistent with Benson and Kirsch's "Problem Denial" phase. In Chapter 4 and following chapters we will consider whether that is the case, and if not, why.

NOTES

1. Anthony Bebbington, ed., *Social Conflict, Economic Development and Extractive Industry: Evidence from Latin America* (London: Routledge ISS Studies in Rural Livelihoods, 2012).

2. Stuart Kirsch, "Afterword: Extractive Conflicts Compared," in *Social Conflict, Economic Development and Extractive Industry: Evidence from Latin America*, ed. Anthony Bebbington (London: Routledge ISS Studies in Rural Livelihoods, 2012), 211.

3. Joshua Karliner, *The Corporate Planet: Ecology and Politics in the Age of Globalization* (San Francisco: Sierra Club Books, 1997), 49.

4. David Korten, *When Corporations Rule the World* (Bloomfield, Connecticut: Kumarian Press, 2001), 22.

5. Jane Kelsey, *Reclaiming the Future: New Zealand and the Global Economy* (Wellington: Bridget Williams e-books, 2015), 34.

6. Ibid., 38.

7. See Peter Benson and Stuart Kirsch, "Capitalism and the Politics of Resignation," *Current Anthropology* 51, 4 (2010): 459–486.

8. David Korten, *When Corporations Rule*, 22.

9. Ibid., 123.

10. Karliner, *Corporate Planet*, 2.

11. Kelsey, *Reclaiming the Future*, 53–54.

12. Ibid., 48.

13. Anthony Giddens, *Runaway World: How Globalization is Reshaping Our Lives* (London: Profile Books, 2002), xxv.

14. Dean Henderson, "The Energy Vampires," *Left Hook* (blog), January 9, 2014, accessed March 24, 2014, www.deanhenderson.wordpress.com.

15. Benson and Kirsch, "Capitalism and the Politics of Resignation."

16. Ibid., 461.

17. See Stuart Kirsch, *Mining Capitalism: The Relationship between Corporations and Their Critics* (Oakland: University of California Press, 2014). Fabiana Li, *Unearthing Conflict: Corporate Mining, Activism, and Expertise in Peru* (Durham: Duke University Press, 2015). Marina Welker, *Enacting the Corporation: An American Mining Firm in Post-Authoritarian Indonesia* (Oakland: University of California Press, 2014).

18. Pierre Bourdieu, *Outline of a Theory of Practice*, trans Richard Nice (Cambridge: Cambridge University Press, 1977), 10 ff.

19. Korten, *When Corporations Rule*, 21–23.

20. Ibid., 123.

21. Tom Wilber, *Under the Surface: Fracking, Fortunes and the Fate of the Marcellus Shale* (Ithaca, New York: Cornell University Press, 2012), 86.

22. David Richardson and Richard Denniss, "Mining the Truth: The Rhetoric and Reality of the Commodities Boom," Paper No 7 (Canberra: The Australian Institute, September 2011), 5–6.

23. "Fossil Fuel Funding to Congress: Industry Influence in the US," Oil Change International, accessed August 12, 2015, http://priceofoil.org/fossil-fuel-industry -influence-in-the-u-s/.

24. Karliner, *Corporate Planet*, 7.

25. Kevin Zeese and Margaret Flowers, "US Climate Bomb is Ticking: What the Gas Industry Doesn't Want You to Know," *Truthout,* March 6, 2013, accessed March 17, 2013, http://www.truth-out.org/news/item/14958-us-climate-bomb-is-ticking -what-the-gas-industry-doesnt-want-you-to-know.

26. Felicity Lawrence and Harry Davies, "Revealed: BP's Close Ties with the British Government," *The Guardian*, May 20, 2015, accessed July 29, 2015.

27. Danny Chivers, "Fracking: the Gathering Storm," *New Internationalist*, December 3, 2013, accessed December 14, 2013.

28. Arthur Neslen, "EU Dropped Climate Policies after BP Threat of Oil Industry 'Exodus'," *The Guardian,* April 20, 2016, accessed June 10, 2016.

29. See Wilber, *Under the Surface.*

30. Karliner, *Corporate Planet*, 41–42.

31. Zeese and Flowers, *Climate Bomb*, 2.

32. Sharon Kelly, "Fracking in Public Forests Leaves Long Trail of Damage, Struggling State Regulators," accessed March 16, 2014, www.desmogblog.com.

33. Karliner, *Corporate Planet,* 76.

34. Presenter A, *Media and Stakeholder Relations Hydraulic Fracturing Initiative 2011*, Houston, Texas, October 31–November 1 (conference presentations posted November 15, 2011, on Texas Sharon's Bluedaze blog, http://www.texassharon.com).

35. "Exxon CEO Talks Arctic Oil Drilling, Risks, Lessons," *New Zealand Herald*, April 3, 2015, accessed May 22, 2015. See also Norwegian Oil and Gas Association, /www.norskoljeoggass.no/en/Facts/Energy-and-climate/.

36. See Richard Heinberg, *Snake Oil: How Fracking's False Promise of Plenty Imperils Our Future* (Santa Rosa: Post-Carbon Institute, 2013). William Engdahl, "America: The New Saudi Arabia?" *Voltaire Network*, March 15, 2013, accessed June 3, 2013,

www.voltairenet.org/article177874.html. Arthur Berman, "IEA Offers No Hope for an Oil-Price Recovery," *Petroleum Truth Report*, November 13, 2015, accessed November 26, 2015, http://www.artberman.com/. Kurt Cobb, "Orwellian Newspeak and the Oil Industry's Fake Abundance Story," July 11, 2014, *Resilience*, accessed October 21, 2014, www.resilience.org. Jeffrey Goodell, "The Big Fracking Bubble: The Scam behind Aubrey McClendon's Gas Boom," *Rolling Stone*, March 1, 2012.

37. See for example *2014 Sustainability Report*, Apache Corporation, accessed January 12, 2015, http://www.apachecorp.com/Resources.

38. R. E. Dunlop and A. M. McCright, "Organized Climate Change Denial," in *The Oxford Handbook of Climate Change and Society*, eds. J. S. Drysek, R. B. Norgaard and D. Schlosberg (Oxford: Oxford University Press, 2011), 144–160.

39. Zeese and Flowers, *Climate Bomb*, 4.

40. Karliner, *Corporate Planet*, 7.

41. For further information, see www.wbcsd.org.

42. Gavin Hilson, "Corporate Social Responsibility in the Extractive Industries: Experiences from Developing Countries," *Resources Policy,* Special Issue 37, 2 (2012): 131–137, accessed December 14, 2013, http://dx.doi.org/10.1016/j.resourpol .2012.01.002.

43. Heledd Jenkins and Natalia Yakovleva, "Corporate Social Responsibility in the Mining Industry: Exploring Trends in Social and Environmental Disclosure," *Journal of Cleaner Production* 14, 3–4 (2006): 272.

44. Heledd Jenkins, "Corporate Social Responsibility and the Mining Industry: Conflicts and Constructs," *Corporate Social Responsibility and Environmental Management* 2, 1 (2004): 23–34.

45. Daniel Franks and Rachel Davis, "The Costs of Conflicts with Local Communities in the Extractive Industry" (paper presented to the First International Seminar on Social Responsibility in Mining, Santiago, Chile, October 19–21, 2011), emphasis added.

46. John Owen and Deanna Kemp, "Social License and Mining: A Critical Perspective," *Resources Policy* 38, 1 (2013): 29–35.

47. Jedrzej George Frynas, "The False Developmental Promise of CSR: Evidence from Multinational Oil Companies," *International Affairs* 81, 3 (May 2005).

48. Emma Gilberthorpe and Glenn Banks, "Development on Whose Terms? CSR Discourse and Social Realities in Papua New Guinea's Extractive Industries Sector," *Resources Policy* 37, 2 (June 2012): 186.

49. Richard Heinberg, *Snake Oil*, 99.

50. Seamus McGraw, *The End of Country: Dispatches from the Frack Zone* (New York: Random House, 2011).

51. Wilber, *Under the Surface.*

52. Zain Shauk and Bradley Olson, "Colorado Drillers Show Sensitive Side to Woo Fracking Foes," *Bloomberg News*, August 28, 2014, accessed November 21, 2014.

53. See "Responsible Development: People," Canadian Association of Petroleum Producers, accessed November 24, 2015, http://www.capp.ca/responsible -development/people.

54. Presenter B, *Media and Stakeholder Relations Hydraulic Fracturing Initiative 2011*, Houston, Texas. October 31–November 1 (conference presentations published November 15, 2011, on Texas Sharon's Bluedaze blog, http://www.texassharon.com).

55. Karliner, *Corporate Planet,* 48–49.

56. Ibid., 48–49.

57. Patricia Widener, "Global Links and Environmental Flows: Oil Disputes in Ecuador," *Global Environmental Politics* 9, 1 (February 2009): 48.

58. "TransCanada's Donation to Town Means Silence on Pipeline," *Bloomberg News*, July 4, 2014, accessed July 7, 2014.

59. K. McNab, et al., *Beyond Voluntarism: The Changing Role of Corporate Social Investment in the Extractives Sector* (Centre for Social Responsibility in Mining, The University of Queensland, Brisbane, 2012), https://www.csrm.uq.edu.au/.

60. Naomi Oreskes and Erik Conway, *Merchants of Doubt: How a Handful of Scientists Obscured the Truth on Issues from Tobacco Smoke to Global Warming* (New York: Bloomsbury Press, 2010), 32.

61. Philip Kitcher, "The Climate Change Debates," *Science* 328, 4 (2010): 1232, accessed August 11, 2014, doi: 10.1126/science.1189312.

62. Oreskes and Conway, *Merchants of Doubt*, 32.

63. Ibid., 19.

64. Dunlop and McCright, "Organized Climate Change Denial," 146.

65. Union of Concerned Scientists, *Toward an Evidence-based Fracking Debate*, Full Report (Cambridge, MA, 2013): 4, www.uscusa.org/HFreport.

66. Richard Berman, "Big Green Radicals: Exposing Environmental Groups" (presentation to the Western Energy Alliance Annual Meeting, Colorado Springs, June 20, 2014), https://s3.amazonaws.com/s3.documentcloud.org/documents/1349204/berman-at-western-energy-alliance-june-20.

67. Presenter C, *Media and Stakeholder Relations Hydraulic Fracturing Initiative 2011*, Houston, Texas. October 31–November 1 (conference presentations published November 15, 2011, on Texas Sharon's Bluedaze blog, http://www.texassharon.com.)

68. Richardson and Denniss, *Mining the Truth*, 7–9.

69. Oreskes and Conway, *Merchants of Doubt*, 18–19.

70. Presenters at *Media and Stakeholder Relations Hydraulic Fracturing Initiative 2011*, Houston, Texas. October 31–November 1 (conference presentations published November 15, 2011, on Texas Sharon's Bluedaze blog, http://www.texassharon.com, emphasis added).

71. Sarah van Gelder, "Why the Corporate Media's Climate Change Censorship Is Only Half the Story," *YES Magazine*, October 11, 2013.

72. Edward Herman and Noam Chomsky, *Manufacturing Consent* (London: The Bodley Head, 2008), xii.

73. Dunlop and McCright, "Organized Climate Change Denial," 146.

74. Zeese and Flowers, "US Climate Bomb," 1.

75. Tom Murphy, "Global Shale Energy: Emerging Technology, Evolving Regulations and Increasing Supply" (presentation to the Business Matters Seminar, School of Business, Auckland University, Auckland, February 27, 2014).

76. Daniel Franks, Jo-Anne Everingham and David Brereton, "Governance Strategies to Manage and Monitor Cumulative Impacts at the Local and Regional Level," ACARP Project C19025 (Centre for Social Responsibility in Mining, University of Queensland, Brisbane, 2012), 4, https://www.csrm.uq.edu.au/, emphasis added.

77. Richard Schiffman, "'Frackademia': How Big Gas Bought Research on Hydraulic Fracturing," *The Guardian*, January 9, 2013.

78. Suzanne Goldberg, "Work of Prominent Climate Change Denier was Funded by Energy Industry," *The Guardian,* February 23, 2015, accessed February 24, 2015.

79. Dunlop and McCright, "Organized Climate Change Denial," 149.

80. Peter Frumhoff and Naomi Oreskes, "Fossil Fuel Firms are Still Bankrolling Climate Denial Lobby Groups," *The Guardian*, March 25, 2015, accessed April 10, 2015.

81. Presenters at *Media and Stakeholder Relations Hydraulic Fracturing Initiative 2011,* Houston, Texas. October 31–November 1 (conference presentations published November 15, 2011, on Texas Sharon's Bluedaze blog http://www.texassharon.com).

82. Korten, *When Corporations Rule*, 150ff.

83. Karliner, *Corporate Planet,* 186–187.

84. "BHP Supports Students in Sustainable Development," *Oil and Gas Financial Journal*, October 10, 2014.

85. See Jonathan Wood, *The Global Anti-Fracking Movement: What It Wants, How Operates and What Next,* Control Risks white paper (October, 2012), cited in Katrina Rabeler, "Gas Industry Report Calls Anti-Fracking Movement a "Highly Effective Campaign," *Yes* Magazine, March 26, 2013, http://www.yesmagazine.org/planet/gas-industry-report-calls-anti-fracking-movement-highly-effective. Also see "Canada's Oil Industry 'in the Middle of a Battle,' Brad Wall Tells Calgary Petroleum Club," *CBC News*, June 8, 2016, accessed June 11, 2016.

86. Dunlop and McCright, "Organized Climate Change Denial," 154.

87. Frumhoff and Oreskes, "Fossil Fuel Firms."

88. "Oil Executive: Military-Style 'Psy Ops' Experience Applied," *CNBC*, November 8, 2011, accessed November 15, 2011.

89. Eric Lipton, "Hard-Nosed Advice from Veteran Lobbyist: 'Win Ugly or Lose Pretty,'" *New York Times,* October 31, 2014, accessed April 9, 2015.

90. Wilber, *Under the Surface,* 188 ff.

91. Adam Federman, "Power Play," *Earth Island Journal*, Spring 2015, 38ff, accessed May 5, 2016, http://go.galegroup.com/ps/i.do?id=GALE%7CA404894276&v=2.1&u=per_mdl&it=r&p=ITOF&sw=w&asid=f79c61ca96cf4d5e211fdfec87294737.

92. Presenter J at *Media and Stakeholder Relations Hydraulic Fracturing Initiative 2011,* Houston, Texas. October 31–November 1 (conference presentations published November 15, 2011, on Texas Sharon's Bluedaze blog http://www.texassharon.com).

93. See Lipton, "Hard-Nosed Advice." It was the hackneyed argument implying that only people who travelled by horseback were entitled to criticize the petroleum industry. A perversion of Jesus' admonition to the scribes and Pharisees: "Let him who is without sin cast the first stone" (John 8:7).

94. "Activist-Turned-Oil-Exec Faces Down Protests," *New Zealand Herald,* April 11, 2015, accessed April 30, 2015. Schuller reportedly participated in environmental

protests while a college student. She left COGA in May 2015 to become strategic advisor for the Natural Gas Initiative at Stanford University, her alma mater.

95. Tisha Schuller, "Deescalating Through the Courts?" COGA CEO's blog, November 18, 2014, accessed April 15, 2015, www.coga.org.

96. Saul Alinsky, *Rules for Radicals* (London: Vintage Books, 1971/1989).

97. "Activist-turned-oil Exec," *New Zealand Herald.*

98. Schuller, "Deescalating Through the Courts?"

99. "Activist-turned-oil Exec," *New Zealand Herald.*

100. Karliner, *Corporate Planet,* 189.

101. Ibid., 191.

102. States such as New York and Vermont have banned fracking.

103. Benson and Kirsch, "Politics of Resignation

104. Maurice Godelier, *Rationality and Irrationality in Economics* (New York: Monthly Review Press, 1972), 258.

105. Thomas Rudel, J. Timmons Roberts and JoAnn Carmin, "Political Economy of the Environment," *Annual Review of Sociology* 37 (August, 2011): 221–238.

106. J. Rogers Hollingsworth and Robert Boyer, eds., *Capitalism: The Embeddedness of Institutions* (Cambridge: Cambridge University Press, 1997), 2.

107. Cited in Rudel, Roberts and Carmin, "Political Economy of the Environment, p 229.

108. Geoff Bertram, "Mining in the New Zealand Economy," *Policy Quarterly* 7, 1 (2011): 13–19.

109. Bertram, "Mining in the New Zealand Economy," 13.

110. Anthony Bebbington, "Extractive Industries, Socio-Environmental Conflicts and Political Economic Transformations in Andean America," In *Social Conflict, Economic Development and Extractive Industry: Evidence from Latin America,* ed. Anthony Bebbington (London: Routledge, 2012), 13, emphasis added.

111. Michael Ross, "Does Oil Hinder Democracy?" *World Politics* 53, 3 (April 2001): 325–361.

112. Kirsch, "Afterword: Extractive Conflicts Compared," 201.

113. Rudel, Roberts and Carmin, "Political Economy of the Environment," 221.

114. See Patricia Widener, *Oil Injustice: Resisting and Conceding a Pipeline in Ecuador* (London: Rowman & Littlefield, 2011). Also see Patricia Widener, "A Protracted Age of Oil: Pipelines, Refineries and Quiet Conflict," *Local Environment* 18, 7 (2013), 841–842, accessed July 26, 2015, http://dx.doi.org/10.1080/13549839.2012.738655.

115. Rudel, Roberts and Carmin, "Political Economy of the Environment," 136.

Chapter Two

The Unconventional Oil and Gas Boom

Since the turn of the millennium, a global boom has occurred in what the petroleum industry calls "unconventional" oil and gas development. Unconventional oil and gas refers to hydrocarbons that cannot be extracted by primary or secondary recovery methods such as *in situ* underground pressure, physical lift, water flood, or water/gas pressure maintenance. Unconventional oil and gas is more expensive to produce and usually occurs in geological formations like shale or tightly packed limestone or sand layers through which oil and gas is dispersed (hence "shale gas" and "tight oil"). Extracting these reserves requires costly equipment, new technologies, and processes such as directional drilling and hydraulic fracturing or "fracking." Fracking involves injecting large volumes of water, sand, and special chemicals under extreme pressure down deep, horizontal wells to penetrate and fracture target geological formations and cause the oil and gas to flow. There are literally hundreds of books, reports, and articles describing the fracking process.[1] The term unconventional oil and gas may also refer to coal seam methane, gas hydrates (deep-sea frozen methane), tar sands, oil shale (not to be confused with shale oil), and even bio-fuels.[2] This study concentrates primarily on land-based exploration and extraction of shale gas and tight oil.

Crucial to understanding the significance of the unconventional oil and gas boom is the Peak Oil controversy, which some industry analysts argue has been put to rest by the unconventional oil and gas boom. In this chapter, after considering both sides of the argument, we will examine the unconventional boom itself particularly in North America and ask whether this boom is all it's reputed to be. We will also examine some of the evidence supporting claims of the harm unconventional oil and gas development causes.

PEAK OIL AND THE OIL GLUT

The notion of "peak oil" implies the world is running out of oil and gas, but there is actually quite a lot of it still remaining. Total proven world oil reserves at the time of writing were estimated at 1,655 billion barrels, up 9 percent over the past five years. The world's gas reserves were estimated to be around 6,972 trillion cubic feet, an increase of 9.5 percent over the same period.[3]

Two problems arise regarding this apparently ample supply of energy (setting aside climate change for the moment). First, allowing for market fluctuations, on current trends demand is expected to outstrip supply over the next quarter- to half-century. Second, the world is running out of *accessible* oil and gas and the rate of commercially viable discoveries is declining. There are doubtless resources that haven't been located yet, but they are likely to be so costly to extract that their price will exceed alternative energy options. Add the issue of climate change and the future for oil and gas starts to look decidedly bleak. The big petroleum corporations know this. Reserves of hydrocarbons will run out sooner or later; it's a matter of when and how fast.

Peak Oil Theory

The theory of Peak Oil was first formulated by geophysicist M. King Hubbert in 1956 and refined by geologist Colin Campbell at the turn of the millennium. Both began with the fact that there is a finite supply of petroleum and sooner or later the world will run out of cheap reserves. As world demand continues to outstrip supply, prices must rise and producers will invent new technologies and processes (like fracking) to try to extract costlier less accessible resources. Prices will become increasingly volatile. Over time supply will inevitably fall, precipitating dramatic declines in production and financial meltdown as companies dependent on cheap hydrocarbon energy fail to adapt and go out of business.

A peak in supply and production doesn't mean the world will suddenly and catastrophically run out of petroleum. It means the supply of easy-to-extract oil and gas reaches a point where it begins an abrupt decline as demand continues, leading E&P (exploration and production) companies to look for reserves that are more costly to extract, resulting in escalating prices and growing competition from alternative energy sources.

The Debate over Peak Oil

No one seriously doubts that peak oil and gas will occur. As Heinberg states, "the timing of peak oil is of great importance. That's what all the fuss is about."[4] Some experts believe the peak has already been reached, while others insist it is still a long way off.[5] Much of the debate centers on definitions of conventional versus unconventional oil and gas, whether unknown resources are included in the projections, and the ultimate scale of the unconventional oil and gas boom.[6]

Besides definitional disagreements, the timing of peak oil is complicated by arguments over:

- the rate of new discoveries,
- calculations of reserves,
- confusion (often intentional) between "proven reserves" and "estimated resources,"
- declining net oil and gas exports,
- the effects of political and financial factors, and
- declining energy return on energy invested from harder to reach reserves– the so-called "net energy cliff."

The consensus among most participants in the debate seems to be that the world has, according to the International Energy Agency (IEA), or will soon reach peak *conventional* oil. And that, despite uncertainties about the additional contribution of unconventional oil and gas, *total peak oil* (i.e., from all sources) is only a few years away—no later than mid-century and probably a decade or two earlier if critics of the IEA's projections are correct.[7]

Why It Matters: Peak Oil and Climate Change

Debates over timing aside, the crucial question is what difference it makes that the world will sooner or later enter an era of expensive, less accessible petroleum. Heinberg maintains that "unless we somehow drastically reduce our dependency on fossil fuels, the impact to the global economy will be serious-to-catastrophic."[8] International agreements and private-sector initiatives to address climate change will only shorten the countdown to peak oil. Heinberg and other catastrophists argue that a perfect storm is brewing with the convergence of four factors: resource depletion, environmental degradation, systematic financial and monetary failures, and peak oil.[9]

As the world moves toward peak oil and petroleum depletion, there will be a redistribution of resource flows and production into energy acquisition, infrastructure, and maintenance. Less and less will be available for consumption, particularly for discretionary consumption.[10] Not everyone will be effected equally, as the Climate Justice movement points out.[11] The realignment of resource and production flow driven by the growth model will lead to a

widening chasm between the rich and the rest, between powerful corporations and wealthy individuals and the increasingly disempowered middle
class and poor. Social conflict will increase. Governments will enter the fray
(and indeed already have) for control of global resources for their people and
nationally based corporations.

Korowicz argues in his treatise, *Tipping Point*, that the key to understanding the consequences of peak oil is to look beyond transport, petrochemicals,
and food production to its wider systemic effects:

> Energy flows and a functioning economy are by necessity highly correlated; our
> basic local needs have become dependent upon a hyper-complex, integrated,
> tightly-coupled global fabric of exchange; our primary infrastructure is depen-
> dent upon the operation of this fabric and global economies of scale; credit is the
> integral part of the fabric of our monetary, economic and trade systems; a credit
> market must collapse in a contracting economy, and so on.[12]

Based on his analysis, Korowicz concludes that an economic, social, and ecological collapse is virtually inevitable unless drastic global policy measures
are taken immediately which he suggests is unlikely.

It is also highly unlikely the world's fossil fuels sector will lead the way
in transitioning to a low-carbon economy to meet the 1.5–2°C cap in global
warming by the end of the century agreed at COP21 (the 2015 UN Climate
Change Conference in Paris). Petroleum transnationals and the IEA still
predict global energy consumption, primarily based on fossil fuels, could
increase by as much as a third over the next half century. According to the
Intergovernmental Panel on Climate Change (IPCC), that would mean a
4–5°C increase which would have disastrous worldwide economic, social,
agricultural, and ecological consequences.

A Temporary Oil Glut

The unconventional oil and gas boom over the past decade, along with stag-
nating demand due to global financial instability, has contributed to an oil
and natural gas glut and a dramatic downturn in world markets. Over the
past five years, gas prices have fallen from a high of over $US 6 per million
BTUs to $US 2.3. Oil prices that were close to $US 120 a barrel in early 2014
plunged to $US 28 a barrel by January 2016 before recovering to $US 50 by
mid-year.[13] Industry analysts point out that even the most efficient frackers
struggle to keep operating at such levels.

As the glut became apparent, unconventional oil and gas producers ramped
up production in competition with OPEC to keep revenue flowing while cutting back on exploration. That led to a 1.6 million barrel per day oil produc-

tion surplus which is likely to increase with the lifting of sanctions on Iran.[14] Even so, pundits do not expect exploration cutbacks to begin affecting production and raising prices until at least 2017.[15] The IEA, World Bank, and the IMF all anticipate demand to pick up slowly and prices to recover gradually over the next five years as reserves peak and the glut clears. On the production side, output from some of the larger US unconventional oil and gas plays like the Bakken, Eagle Ford, and Haynesville are expected to decline substantially over the same period.[16] In sum, the glut in world oil and gas supplies is expected to be temporary as demand picks up and prices rebound.

THE UPSIDES AND DOWNSIDES OF THE UNCONVENTIONAL OIL AND GAS BOOM

The boom in unconventional shale gas and tight oil started around 2000. By 2012, the US Energy Information Administration (EIA) calculated that directional drilling and shale fracking was producing 14 percent of the world's natural gas supply and more than 20 percent of the US gas supply. There were 493,000 active natural gas wells in the US in 2009, around 90 percent of which were developed with hydraulic fracking. The tight oil boom is no less impressive. Approximately 50 percent of US oil production came from fracking in 2015.[17] By 2019, the EIA expects the unconventional oil supply outside the United States to reach 650,000 barrels per day, while US output is forecast to roughly double to 5.0 million barrels per day.[18] In Texas's Barnett Shale, one of the premier plays in the United States, more than 11,000 wells were drilled from 2003-2010. A majority of those were within the Dallas-Ft. Worth metropolitan area which sits on big reserves of gas and oil.

The Claimed Benefits of Unconventional Oil and Gas

Besides energy security,[19] the petroleum industry claims the unconventional oil and gas boom can boost a country or region's GDP through availability of cheaper energy, royalties, taxes, and export earnings. The United States, held up as a prime example, is on course to become the world's foremost petroleum producer, surpassing Saudi Arabia, leading some media commentators to coin the term "Saudi America."[20] Consumers will also benefit from lower gas utility bills, and industrial customers will benefit by switching to cheaper fuels.

Shale gas and tight oil exploration, site preparation, drilling, and production are also said to generate employment growth to a local economy. Testifying before the House of Representatives Subcommittee on Energy and Power in February 2013, energy expert Daniel Yergin claimed the boom would produce up to three million jobs in the United States by 2020.[21]

For larger industrial nations, a boom in unconventional oil and gas can lead to increased geopolitical influence on the world stage. A surge of tight oil supplies onto the world market in recent years has led to a price war between OPEC producers and North America. Russia has been able to exert pressure on European countries and deflect some of the effects of internal sanctions for its proxy invasion of the Ukraine through its control of gas supplies to these countries. The United States is aiming to counterbalance that influence by gearing up to export liquefied natural gas (LNG) to its European allies.

The industry claims the environment and world climate will benefit from the unconventional oil and gas boom. Shale gas is said to be a "transition fuel" that will help the world mitigate the effects of climate change. In the United States, for example, the shift from coal to gas in the electricity sector is claimed to have contributed to a significant reduction in CO_2 emissions; CO_2 emissions from burning coal are more than double those produced using gas. The industry claims horizontal drilling, fracking, and related technologies are not harmful to the environment or humans. Companies use safe, responsible operating procedures "to minimize risks, ensure environmental stewardship, and efficiently recover energy resources."[22]

In spite of such sanguine claims there are growing doubts about whether the unconventional oil and gas boom is all it's cracked up to be. Critics argue the fossil fuel-dependent world economy is ensnared in a series of destructive feedback loops driven by the growth paradigm which the boom is exacerbating. The oil and gas boom depends upon increasing demand to maintain prices and make it commercially viable to invest huge amounts of capital in exploration and infrastructure development. This in turn is likely to hasten the arrival of "peak oil" and, by pouring increased amounts of carbon, methane, and other gasses into the atmosphere, accelerate climate change which in turn is predicted to impact food production, economic stability, social well-being and ecosystems.

All of this means that alarm bells are beginning to ring regarding the unconventional oil and gas boom. Critics point to the following warning signals.

A Boom Heading for a Bust: Rapidly Depleting Wells

A number of respected industry analysts are predicting an eventual bust in the unconventional petroleum boom. One key reason is that unconventional oil and gas wells deplete at much faster rates than conventional wells. E&P companies have to continually pour in capital and drill more wells just to maintain production at acceptable levels, particularly where they have already tapped the best areas or sweet spots. The remaining deposits are less productive and more costly to locate and drill.

What does the evidence indicate? Based on data from unconventional fields across the United States, production from shale gas and tight oil wells declines on average around 90 percent after four years.[23] Some wells decline as much as 60–80 percent in the first year. The average depletion rate in the Bakken Shale is 69 percent in year one and 39 percent of remaining reserves in year two. In light of high depletion rates the number of new wells required annually to keep up with rising demand will outstrip production by 2019 and will be 30,000 wells a year short by 2040. Unconventional drilling overall must increase by 77 percent above 2010 levels if energy production is to keep up with growing demand.[24] That's presuming there are enough reserves and drilling locations in the United States (and "frontiers" like New Zealand) to accommodate the massive number of new wells needed.

Rising Costs and Declining EROEI

The petroleum industry maintains that new technologies and processes are reducing costs, making it more attractive to drill for those who are sufficiently cashed up, even during the current market slump. On the downside, unconventional oil and gas are more costly to produce than conventional petroleum, anywhere from $US 30 to $US 80. It's no wonder more exploration and production companies are going to the wall.

In addition, EROEI or "energy returned on energy invested" is declining fast in today's shale gas and tight oil fields. EROEI is the net amount of energy yielded by a particular resource, such as oil from a horizontally drilled and fracked well. In other words the difference between how much energy it takes to extract a resource and the amount of energy in the final product.[25] The lower the EROEI, the higher the production costs.

In the 1930s, one barrel of oil's worth of energy invested in exploration and production produced a hundred barrels of oil. Today the ratio of energy yield to energy investment is closer to 10:1.[26] Even with efficiency gains from new technology, unconventional wells in fields of declining quality require more energy to drill. Once EROEI declines past the 10:1 ratio it goes over what analysts call an "EROEI cliff" where the energy input required to produce a barrel of energy increases exponentially. In spite of claims about the economic benefits of tight oil, shale gas, and tar sands, each of these fuel sources has an EROEI of 5:1 or less![27] In summary total production may be impressive but most wells, at least in the United States, are unprofitable, and companies are scrambling to keep afloat.

Exaggerated Energy Reserves

The petroleum industry has a history of deliberately over-estimating reserves and over-promising on economic benefits, particularly with respect to unconventional oil and gas.[28] Arthur Berman, a leading geological consultant, estimates that US shale gas reserves have been overstated by at least 100 percent.[29] Instead of the much-touted one hundred years supply of gas, Berman calculates there are no more than twenty years of reserves left. The same exaggerated claims have been made about oil reserves, according to Cobb.[30]

A frequently used tactic employed by industry representatives, media oil boosters, and government bureaucrats alike is to blur the distinction between "reserves" and "resources." There are important differences in how these terms are defined that are crucial for how geological analysts estimate the amount of petroleum product in any given basin or play. We'll discuss these terms in more detail in a later chapter. The public and media generally aren't aware of the technical differences and so don't know that the terms are being manipulated to support PR hype about the huge potential of the unconventional boom.

Exaggerated Job Claims

Independent research suggests industry job estimates side-step the question of how many lasting, well-paying petroleum exploration and production jobs are generated for local residents. Typically most jobs other than site preparation go to experienced oil workers from outside the region, at least until the industry becomes well-established. Furthermore, the notion of "indirect jobs" is used to pad employment numbers. Claiming trucking and engineering work are part of the petroleum industry is one thing, but including sales clerks and fast-food servers stretches credibility (sex workers are seldom mentioned). Employment generation is touted by the industry and local councils as a major selling point for establishing or expanding the industry.

A Giant Ponzi Scheme?

As noted earlier, many exploration and production companies have gone out of business during the most recent unconventional oil and gas boom. For most companies, operating costs and debt have outstripped earnings, and the gap is getting wider annually. The current downturn in the global oil market is forcing companies who until now managed to keep production levels up by drilling more wells to cease exploration, sell productive assets, borrow more capital, and defer tax payments.

Analysts have described shale gas and tight oil production as a treadmill: more wells to maintain production to keep shareholders happy, leading to more costs that outstrip earnings, requiring more capital borrowing to drill more wells. Some are now calling it a "giant Ponzi scheme" because these companies are only able to keep afloat by duping unsuspecting investors.[31] In 2013 it cost approximately $US 42 billion a year in the United States to drill 7,000 new unconventional gas wells to main production at current levels. The return to drilling companies in gas produced was $US 32.5 billion, a $US 9.5 billion deficit in one year. Even after attracting new investors, some companies have been saddled with billions of dollars of debt. No wonder business commentator SR Srocco claimed that "fracking shale is destroying oil and gas companies' balance sheets."[32]

Problems with Natural Gas as a Transition Fuel

Recent research has raised serious doubts over industry claims that natural gas is an ideal "transition fuel" to help the world move to a low-carbon economy. The Fifth IPCC report[33] acknowledged that natural gas may have a role because it releases fewer carbon emissions than coal or oil; but the panel also raised questions about its feasibility and long-term efficacy without advanced carbon-storage technology. Clark[34] points out that the notion of gas leading a transition to alternative sources is a myth, because industry experience shows it takes decades for companies to recover capital investments in a new developments, repay loans, and start earning profits to satisfy investors. Petroleum companies would not embark on such a major undertaking to build the necessary infrastructure, and convert gas to other applications without extensive government (taxpayer) support.

Peer-reviewed scientific studies have also documented the large amount of fugitive methane released in the process of extracting, flaring, and distributing unconventional natural gas. Methane has been found to be ten times more polluting as a greenhouse gas than CO_2. Combined with the CO_2 generated by *burning* natural gas, an average leakage rate of 3 percent from *producing* it would make natural gas more polluting than coal. That rate is regularly exceeded in observational studies. A team of Cornell University researchers headed by Anthony Ingraffea and Robert Howarth found that 3.6 percent to 7.9 percent of the methane from shale-gas production escapes to the atmosphere in venting and leaks over the life-time of a well.[35] At that rate the world will reach $1.5°C$ warming in fifteen years or less, after which dangerous tipping points in the earth's climatic system begin to occur.[36] Leakage of ethane, a chemical cousin of methane, is having similar effects.[37]

Chivers sites the Fifth IPCC report and international scientific studies that indicate if we wish to limit global warming to 2°C we can only use a quarter of *known* oil and gas reserves and we must stop burning fossil fuels by 2050. More recent studies suggest the figure for total resources may be closer to seventy-five percent. "There is no space at all in this equation for unconventional fuels; their extraction is quite simply incompatible with maintaining a livable climate."[38]

The Carbon Bubble

Shareholders and financial regulators are beginning to ask hard questions of petroleum corporations about what happens if stringent measures are introduced to address climate change that end up restricting fossil fuel production. Energy analysts have coined the term "carbon bubble" (or "stranded assets") for this scenario, referring to how the value of companies who produce fossil fuels or depend on them could suddenly collapse. Current company valuations assume that all known fossil fuel reserves will be available for consumption at prices set by the market. If use of these reserves were restricted by international agreements or artificially priced via carbon markets to make them unaffordable in comparison to renewables, the value of these companies would plummet. Potential losses to fossil fuel companies have been estimated at anywhere between $US 28 trillion and $US 100 trillion over the next two decades.[39]

This is not just doomsayer histrionics. The United Kingdom's Committee on Climate Change has warned the carbon bubble poses a serious threat to the UK economy. Other UK financiers and academics have made similar statements.[40] The European Systemic Risk Board, established after the 2008 financial crash, has warned in a new report of economic "contagion" if moves to a low-carbon economy happen too late and the changes are too abrupt.[41] Shareholders rightly are beginning to worry their investments could be worthless.

THE HARMFUL EFFECTS OF UNCONVENTIONAL OIL AND GAS DEVELOPMENT

The following discussion briefly highlights findings from peer-reviewed scientific research, government studies, media investigations, and even industry reports about the harmful effects that are associated with unconventional oil and gas operations. These are the sorts of effects that government legislation and supposedly rigorous regulatory systems are set up to monitor and "mitigate," if not prevent. They are also the sorts of effects that politicians, ministers, policy advisors, and local councils are expected to be aware of and address. Germany, France, Tunisia, Wales, Scotland, New York, Maryland, and numerous other jurisdictions across America, Australia, and Canada have considered fracking so dangerous they have passed moratoria or banned it altogether. The focus in the following discussion is on unconventional shale gas and tight oil operations in their entirety, not just fracking.

Referring to operations "in their entirety" requires further clarification. Benson and Kirsch argue that the business operations of industries with a reputation for causing harm are predicated on their ability to avoid liability for most of this harm. To expose such corporate maneuvers, they recommend "a focus on the concrete biological, social, and environmental problems caused by corporate capital and the tactics and strategies that corporations pursue to avoid or manage the resulting liabilities."[42] For clarity, the following discussion does not include the business planning that goes into assessing (a) the potential negative impacts of developing a given well or field, (b) the techniques a company is prepared to utilize to manage or "mitigate" these impacts, and (c) the stringency of the government's regulatory regime and the extent to which the company is able to externalize or avoid paying for these harmful effects. Nor on the strategies and tactics that petroleum companies

Table 2.1.   The Negative Impacts of Unconventional Oil and Gas Development.

Economic	• Labor market disruption, exaggerated local employment claims
	• Damage to commodity export & tourist sectors
	• Infrastructure costs
	• Impacts on housing availability, rents, and land values
	• Pressure on services and civil administration
	• Non-renewable resource depletion
	• Macro-economic impacts
Environmental and Health	• Evidence of shale gas and CSG methane leakage exacerbating climate change
	• Surface and groundwater contamination
	• Waste treatment problems
	• Air pollution and health problems
	• Increased seismicity (earthquakes)
Social	• Municipal government overload, services breakdown
	• Dislocation, homelessness
	• Increased crime
	• Social conflict, widening disparities
	• Traffic congestion, accidents
	• Noise pollution
Cultural	• Detrimental effects on indigenous peoples' unity, values and customary practices
	• Heritage and amenity destruction
	• Disruption to community identity and networks

employ to portray their operations and mitigation plans as "best practice," promote their activities to the public, and defend themselves from criticism and disruption. We will examine such planning and strategizing later in this book in relation to the petroleum industry's efforts to expand their operations in New Zealand.

Based on a scan of international studies, reports and institutional websites, the following frequently occur as a consequence of unconventional oil and gas exploration and production (table 2.1).

Let us consider a few recent findings about these harmful effects.

Economic Impacts

Governments and petroleum companies emphasize the economic benefits of oil and gas exploration and production in order to justify such developments to the wider public. But research shows there are potential economic downsides to land-based unconventional oil and gas activities which are often ignored or glossed over by politicians and corporate spokespersons.

a. A Mixed Story for National and Local Development

Petroleum exploration and production has the potential to make significant contributions to national and regional GDP, at least in the short term. Much depends on the amount of available reserves, the level of royalties and taxes that governments negotiate with the industry, and the extent to which the financial benefits are able to be captured by regions and communities.

Development is not only a distributive question but one of pace and capacity. The reality, as numerous studies testify, is that countries or regions dependent on extractive industry for development are subject to boom and bust cycles and the economic devastation that accompanies it. A December 2015 report by the Federal Reserve Bank of Philadelphia found that North Dakota, one of the highest performing US states over the past five years due to unconventional oil production from the Bakken Shale, experienced the largest economic downturn in 2015. The state government was facing a $US1.1 billion shortfall in tax revenue in 2016.

Jacquet[43] examined data from a range of studies of communities affected by shale gas development in Wyoming and Pennsylvania, and concluded:

> Some sectors or communities will benefit much more than others. Businesses or residents not directly tied to the energy industry may have to deal with inflationary or employment pressures while not seeing gains in revenue. Job growth can be stratified, as while new jobs will be created, not all workers will be suited for or interested in these jobs. Expectations for economic benefits are

often unrealistically high, and while economic and job growth does occur, these expectations are not met. [44]

Furthermore, the *cumulative impacts* of unconventional oil and gas on local communities and environments are often overlooked by governments and municipalities when championing the establishment of the industry. Christopherson[45] and her team reviewed a range of reports on the economic effects of gas development in the Marcellus Shale and concluded that the faster the pace of a project, the higher the cumulative costs to the community and region. When the boom is over, the community may lose population and is still left to pay for infrastructure built for the industry. Barth[46] reviewed several studies of the economic effects of unconventional oil and gas development and concluded that areas that once had thriving extractive industries that declined ended up suffering long-term unemployment; and counties that placed all their chips on energy development underperformed economically compared with similar counties that had little energy development or were more diversified.

There is often a downside for non-extractive sectors in regions and communities impacted by oil and gas development. Christopherson[47] points to the "crowding out" of certain businesses as an example of negative regional development. A literature review by members of Catskill Mountainkeeper[48] concluded that "existing industries such as tourism, outdoor recreation and agriculture are incompatible with, and threatened by, gas drilling." Greer, Marquet, and Saunders[49] observed that while there are opportunities for businesses to piggy-back off the oil and gas development, there could be negative effects on sectors dependent on export such as agriculture, forestry, and horticulture due to a rising exchange rate and competition for labor.

b. Dubious Employment Growth

As noted earlier, it is not unusual for mining and petroleum companies, trade organizations, and government agencies to exaggerate job numbers from oil and gas development. Richardson and Denniss reviewed Australian industry figures and accused extractive industry representatives and economic consultants of "spinning the truth."[50] Barth found that a report by the Marcellus Shale Coalition, an industry-funded organization, overstated jobs created by the gas industry in Pennsylvania by 25 percent. Half of these numbers were "indirect employment": health, leisure, hospitality, and education.[51]

c. Infrastructure Impacts

When unconventional oil or gas development takes off, communities are usually left with most of the cost of repairing and upgrading roads and building new infrastructure. Cost sharing varies from state to state and country to country, but the industry rarely shoulders more than a small portion of the burden. A 2011 New York Department of Transport study estimated the state's bill to repair roads and bridges at up to $378 million. A 2012 Texas state report calculated that truck traffic associated with oil and gas development "has caused an estimated $2 billion in damage to Texas roads, and there was no revenue source to pay to fix it."

Communities are also impacted by traffic congestion and accidents from a rapid build-up in truck traffic. Truck traffic can vary widely, depending on site location and characteristics, available equipment and other circumstances. The EPA estimates that it takes on average around 200 truck trips to transport a million gallons of fresh or waste water. If the average fracked well consumes between 2 and 10 million gallons that equates to somewhere between 400 and 2000 truck trips per well. Other studies suggest these figures are extremely conservative.

d. Overwhelmed Municipal Government and Services

A boom in unconventional oil and gas can place considerable pressure on local services as well as on the administrative and financial resources of local councils. Much depends on the scale and pace of the development, and on the existing capacities of the community. Larger municipalities are usually in a better position, at least initially, to cope than isolated rural towns.

Social impact studies show that communities can expect increased demands on schools, health care, hospitals, environmental monitoring/remediation, and ambulance and fire services (e.g., chemical spill, fires, accidents on rigs). Most municipalities lack the specialist equipment needed and have to obtain these at an extra cost. There are likely to be greatly increased pressures on public administrative services such as planning and zoning, permitting, assessments, and housing assistance. Law enforcement authorities find themselves having to cope with increased demands from traffic enforcement, crime, and drug use.[52]

e. Rents and Property Values

A review by Catskill Mountainkeeper[53] noted that while some (not all) property owners and speculators may benefit from escalating property prices and rents, people who are not employed in the petroleum sector or who are living on fixed incomes may find themselves squeezed out of the housing market. Boomtowns typically are characterized by an increase in homelessness and rising demands on social services.

Some property owners do decidedly worse when a mining or drilling boom begins. A lot depends on proximity to extractive operations. According to

Dutzik, Ridlington, and Rumpler,[54] "fracking can reduce the value of nearby properties as the result of both actual pollution and the stigma that may come from proximity to industrial operations and the potential for future impacts." The US insurance industry has begun refusing coverage on properties located near unconventional oil and gas wells because of "the unique risks associated with the fracking process."[55] In Australia farmers whose property has been affected by coal seam gas (CSG) development receive little in compensation and struggle to sell their properties.[56]

Environmental and Health Impacts

A considerable body of scientific research and regulatory reports has documented harmful environmental and health impacts associated with unconventional oil and gas operations. A report for the European Commission Directorate General for the Environment on unconventional petroleum production in North American and Europe[57] based on a literature review, operational reports, specialist technical assessments, and site visits identified the following risks to human health and the environment (table 2.2).

Let us examine more closely some of the evidence linking unconventional oil and gas production to environmental and health effects.

a. High Water Demand

Global freshwater supplies are under increasing pressure from world population growth, agriculture, and industrial expansion exacerbated by climate change. By 2050, more than 40 percent of the global population may be living in areas of severe water stress. Worldwide water demand is set to increase

Table 2.2. Preliminary Risk Assessment across All Phases of Unconventional Production.

Environmental Aspect	Individual Site	Cumulative Operation (per field)
Groundwater contamination	High	High
Surface water contamination	High	High
Water resources	Moderate	High
Release to air	Moderate	High
Risk to biodiversity	Moderate	High
Noise impacts	Moderate-High	High
Visual impact	Low-Moderate	Moderate
Seismicity	Low	Low
Traffic	Moderate	High

Source: Adapted from Mark Broomfield, Report to the European Commission Directorate-General Environment, Table ES1 (2012), vi.

55 percent by mid-century. A UN report[58] notes that water use for energy production will increase 20 percent, particularly from unconventional oil and gas because it is more water intensive than conventional production. [59] Compounding the problem, nearly half the wells fracked in the US since 2011 are located in regions with high or extremely high water stress.

The amount of water needed per well varies depending on the type of geological formation being drilled, its depth, and porosity. A review of fifteen reports and journal articles[60] on North American unconventional oil and gas water use estimated that water demand for drilling and fracking over the lifetime of a well ranges between 8.4 million liters and 28.2 million liters per well. These are conservative estimates; some projects on record have used as much as 40–50 million liters per well, and most drill pads have multiple wells. [61]

b. Surface and Groundwater Contamination

A popular mantra of the oil and gas industry is that there's never been a confirmed case of groundwater (e.g., drinking water) contamination in thousands or millions of fracked wells.[62] The mantra always contains the qualifier "confirmed," because the industry challenges virtually every monitoring report and scientific finding on drinking water contamination. The problem with the mantra is that it is simply wrong, as documented by numerous investigations by independent researchers and regulatory agencies.[63] A Duke University study found that households within a kilometer of Marcellus Shale gas wells were at higher risk of having their drinking water contaminated by gases. Methane concentrations in the water were six times higher and ethane twenty-three times higher for homes within a kilometer or less of a shale gas well. The study confirmed previous research[64] that the gas contaminating property owners' water was likely due to poor well construction, though fracking effects could not be ruled out. Ingraffea and his colleagues[65] reviewed data from 41,000 conventional and unconventional oil and gas wells drilled in Pennsylvania between 2000 and 2012, and found a six-fold higher incidence of cement and/or casing impairment (failure) for shale gas wells compared with conventional wells. Similar findings were reported by Davies[66] and by Campbell and Horne.[67]

Contamination of aquifers, drinking water, and surface waterways can also occur through spills and well blowouts, failure of waste pit linings, dumping of wastewater into rivers, and during transport and treatment.[68] In Alberta in 2009 oil and gas companies spilled 23.3 million liters of wastewater.[69] In the United States between 2009 and 2014 there were more than 21,651 *reported* spills involving 175 million gallons of wastewater from ruptured pipes, overflowing storage tanks, accidents, and deliberate dumping.[70] Even in New

Zealand, with its small-scale industry, there were 363 reported oil and waste spills between October 2011 and July 2015.[71]

c. Waste Treatment Problems

Most municipal waste treatment facilities, assuming they are close enough to the oil and gas fields, lack the technical capability to safely process all the chemicals, toxins, and radiation in water produced from unconventional oil and gas production.[72] This waste needs to be processed in specialized industrial waste facilities which are costly and rarely available in remote regions. There are numerous reports of pollution of waterways and sediments downstream from municipal plants in the Marcellus Shale and elsewhere.[73] In 2010 Pennsylvania well operators reported recycling about half of their 2.8 million liters of produced water; the rest went to plants that simply discharged the inadequately treated effluent into rivers.[74]

d. Air Pollution

Studies indicate that unconventional oil and gas production is a serious source of air pollution and related health problems in areas that experience large amounts of drilling.[75] One of the most comprehensive reviews of the health hazards of shale gas and tight oil production was published in 2014 by Physicians Scientists & Engineers for Healthy Energy (PSE), the University of California, Berkeley, and Weill Cornell Medical College.[76]

In 2009, Wyoming failed to meet federal standards for air quality for the first time in its history. In sparsely populated Sublette County, which has some of the highest concentrations of Wyoming's unconventional oil and gas wells, "vapors reacting to sunlight have contributed to levels of ozone higher than those recorded in Houston and Los Angeles."[77] "A 2009 study for the US Environmental Defense Fund in five Dallas-Fort Worth area counties experiencing heavy Barnett Shale drilling activity found that oil and gas production was a larger source of smog-forming emissions than cars and trucks."[78]

Research by the National Center for Atmospheric Research in Boulder, Colorado, found after three years of monitoring one natural gas field that there was a potentially toxic build-up of petroleum hydrocarbons (benzene, toluene, etc.) in the area. After an epidemiological survey, the study concluded that residents living less than half a mile from natural gas wells were at greater risk of negative health effects than residents living more than half a mile away.[79] Food and Water Watch cited two peer-reviewed Colorado studies, one of which tested air quality near fracking operations and detected harmful levels of various toxic solvents; a number of these solvents affect the endocrine system.[80] The second study found that residents living close to drilling and fracking operations had an increased cancer risk, primarily due to potential exposure to benzene.[81]

e. Seismicity (Earthquakes)

There is mounting evidence of a direct link between earthquakes and unconventional oil and gas operations, particularly recycling of wastewater by injection into depleted wells. A study in the Proceedings of the National Academy of Sciences analyzed sixty-seven earthquakes recorded between November 2009 and September 2011 in a 43.5-mile area of the northern Texas's Barnett Shale formation.[82] The study found that "all twenty-four of the earthquakes with the most reliably located epicenters originated within two miles of one or more injection wells for fracking wastewater disposal." Similarly, William Ellsworth, a scientist at the Earthquake Science Centre in the US Geological Survey,[83] found a striking correlation between the incidence of earthquakes in Middle America, the take-off of the fracking boom after 2000, and increased wastewater injection from fracking.[84] Van der Elst and his team[85] examined the phenomenon of increased seismicity in Middle America and concluded injection of drilling wastes was "driving faults to their tipping point."[86] Wells that disposed of very large volumes of water over an extended time seemed most likely to trigger larger earthquakes of magnitude four or five.[87]

Social and Cultural Impacts on Communities and Indigenous Peoples

Human interactions with the environment are not simply about the destructive impacts that individuals, corporations, and governments have on nature but the consequences of that damage for the well-being of communities, families, and individuals.[88] When petroleum or mineral resources are discovered in commercial quantities, the process of initiating a project and exploiting a field develops a momentum of its own and can have far-reaching social consequences for nearby communities and indigenous nations.

In the 1970s and 1980s, social researchers studying the impacts of extractive industries on communities developed what has become known as the boomtown model. Heinberg observes that the boomtown syndrome "tends to characterize unconventional fracking operations more than conventional oil and gas development, because shale gas and tight oil per-well production rates tend to decline so steeply (making the boom briefer and the decline steeper), and also because damage to the environment and to local roads, public health, and community solidarity can be much more serious."[89]

The model was initially developed to try to account for what was happening to rural communities and how these communities responded to the impacts. Boomtown developments occur in stages and usually evolve over extended boom-and-bust cycles. For rural communities resource booms often mean the industrialization of what was once a sparsely populated agricultural and/or natural landscape. The boom results in cumulative impacts over time that can reach tipping points triggering major changes in social, environmental, and economic systems.[90] Recent research has found that the picture is more complicated than first portrayed, particularly taking account of urban/rural differences, but the model is still instructive.

The main social impacts identified by the boomtown model are:

- Social disruption, family breakdown, increased disparities and conflict (for example, the elderly, fixed income, unskilled, disabled are marginalized, experience accommodation problems, and/or are forced to move out)[91]
- Increased crime, substance abuse, mental health problems
- Pressures on community institutions, services, and accommodation
- Undermining of local culture, community identity, and lifestyles
- Impacts on indigenous peoples' sovereignty, culture, and unity[92]

Comparative research by Widener[93] on the effects of petroleum developments on communities in North and South America identified eight common experiences:

1. Most such developments are imposed on communities, and they aren't in a position to say "No." From the perspective of national governments and the industry, the communities are already at the margins of the democratic process. The decisions have already been made.
2. Communities are forced to respond when the oil and gas industry comes into the area, with little time to consider or organize a response.
3. Communities and organizations have less information and knowledge than government or industry. They may lack access or are simply unable to understand or cope with the flood of reports, documents, studies, and media releases. It is difficult for them to undertake their own baseline studies and analysis because of a lack of resources, expertise, and time.
4. Community leaders often feel unprepared, vulnerable, and overwhelmed by demands placed on them compared with national leaders. Some feel stressed, threatened, paranoid, or harassed not just by industry advocates but by erstwhile neighbors with conflicting views.
5. Communities are changed by the impacts of petroleum development, for better or worse depending on one's viewpoint. There are many unintended

consequences which local authorities and companies often attempt to minimize or claim to "mitigate."

6. Local environmental protection standards and social values are often at odds with those of the government and petroleum corporations. Community identity, culture, lifestyles, and social fabric are often overwhelmed by outside influences triggering a deep sense of loss and anomie.
7. Petroleum companies bring change and strangeness. They have different ways of doing things, personnel from outside and a high staff turnover which contribute to changing populations.
8. Communities don't respond uniformly. Not all local residents choose to engage or are equally concerned. Others are highly motivated to act, but can be overwhelmed by the issues and demands on their time. Divisions occur within the community, and stresses arise within families and among friends.

A scan of rural community studies over the past couple of decades[94] suggests that community responses to resource booms tend to fall into distinctive phases: initial enthusiasm, uncertainty, near panic, and eventual adaptation (or people choose to leave).

Implications for New Zealand

Unconventional onshore oil and gas production in New Zealand is tiny by world standards, relatively recent in its application and concentrated in the Taranaki region. In 2011, a Taranaki Regional Council report found there had been only forty-three fracking events in twenty-eight wells with no evidence of environmental problems. Since then, there have been perhaps two or three times that number of fracks. With the exception of a handful to rural townships, there have been few of the economic, social or environmental effects experienced by overseas communities caught up in the unconventional oil and gas boom.

Petroleum industry plans to expand unconventional onshore exploration and production in New Zealand have been delayed by the global oil market slump, but companies who have weathered the storm are already gearing up to expand their operations. Certainly the Government and industry are hoping there will eventually be a "game changing" boom. At the same time, the Parliamentary Commissioner for the Environment has raised concerns over the environmental impacts of unconventional oil and gas development based on overseas evidence and pointed to gaps in New Zealand's regulatory regime, while urging the Government to give greater priority to dealing with climate change.[95]

CONCLUSION

The boom in shale gas and tight oil production has revolutionized the petroleum industry, turned world markets on their heads, and has important ramifications for the Peak Oil debate. On the surface, it appears the unconventional boom has the potential to delay a dramatic decline in petroleum production for decades. Countries and regions have benefited from lower petroleum prices (aided by consumption subsidies), taxes and royalties, energy security, and modest employment growth. Industry advocates even claim the boom in shale gas could play a vital role in transitioning to an alternative-energy economy.

However, it is becoming increasingly clear that the unconventional oil and gas boom is not all it's cracked up to be. Not only are there doubts over its longevity, but evidence suggests the benefits have been overstated, the costs to society and the climate understated, the volume of reserves exaggerated, and the longevity of the boom is doubtful. First, average output per well declines at a much faster rate than convention well, and companies have to pump more and more wells just to maintain current production. Second, costs of production continually outstrip gains meaning companies must take on debt and entice investors in order to fill the gap, giving the appearance of a giant Ponzi scheme to the whole fracking enterprise. Looming over the industry is an emerging consensus among climate scientists, international policy makers, economists and many business leaders in the wake of a series of IPCC reports and COP21 that urgent constraints are needed on production and consumption of fossil fuels if climate warming is to be kept under 2°C to avoid catastrophe. Suddenly the insurance industry, financial regulators, shareholders, and even corporate boards are beginning to talk about "stranded assets" or a "carbon bubble" that could transform petroleum exploration and production into a sunset industry virtually overnight.

Mounting evidence suggests that unconventional oil and gas production has become one of the world's major harm industries, reliant on externalizing most of the non-operational costs it generates in order to profit and grow. It is a major contributor to climate change, damages regional and national economies, pollutes local resources and destroys environments, causes numerous health problems, and disrupts the social organization and cultural fabric of communities and indigenous peoples. To persist and grow as a sector, the petroleum industry has had to employ a portfolio of strategies, often in collaboration with supportive governments, to justify the existence of the industry and defend itself against criticism and organized protest. The industry has been forced to devote more time and resources to challenging reputable scientific findings with its own counter-research, questioning the credibility of established scholars and knowledgeable environmentalists, and generating "reasoned discussion" (from a privileged position of technical expertise) about the industry's continued role in the world's future energy mix. Such a fabricated debate serves to legitimate the continued existence of the industry, justify the world's dependency on fossil fuels, and deflect attention of the "Silent Majority" from the harm the industry is doing.

In the next few chapters we will explore efforts by the New Zealand government to encourage the expansion of oil and gas development. As a "frontier" region during a global unconventional petroleum boom at a time when public as well as scientific concerns are escalating over climate change, New Zealand provides an opportunity to examine how international petroleum corporations in conjunction with the state have tailored their strategies to take account not only of a changing global energy scene but of the particular social, cultural and historical circumstances of the country. And how environmental organizations, community groups and indigenous Māori have utilized their national and global networks to gather evidence, learn about and respond to petroleum industry strategies and "sustainable growth-driven" government manoeuvers.

NOTES

1. See Michael McElroy and Xi Lu, "Fracking's Future: Natural Gas, the Economy and America's Energy Prospects," *Harvard Magazine*, January–February 2013, 24–27. Tony Dutzik, Elizabeth Ridlington, and John Rumpler, "The Costs of Fracking: The Price Tag of Dirty Drilling's Environmental Damage," Environment New York Research & Policy Center, Fall 2012, accessed February 1, 2013, www .environmentnewyorkcenter.org/. Danny Chivers, "Fracking: The Gathering Storm."

2. For a discussion of different unconventional sources see David Hughes, *Drill Baby Drill: Can Unconventional Fuels Usher in a New Era of Energy Abundance?* (Santa Rosa: Post Carbon Institute, 2013), 48–52.

3. Energy Information Administration, "International Energy Statistics" (EIA, Washington, DC, August 22, 2015), http://www.eia.gov/cfapps/ipdbproject/.

4. Richard Heinberg, *Snake Oil*, 22.

5. Some industry commentators suggest major petroleum corporations are betting that most countries will not implement measures to *dramatically* curtail oil and gas extraction and production or cut carbon emissions, regardless of international COP12 commitments.

6. See Bruce Darrell, "Why Confusion Exists over When Peak Oil Will Occur," Foundation for the Economics of Sustainability, January 12, 2007, accessed June 7, 2014, http://www.feasta.org/2007/01/12/why-confusion-exists-over-when-the-oil -peak-will-occur/. Also Richard Heinberg, *The End of Growth: Adapting to Our New Economic Reality* (Gabriola Island, BC: New Society Publishers, 2011), 19 ff.

7. David Knight, "Climate Change and Peak Oil: Two Sides of the Same Coin?" (paper presented to a public symposium organized by Winchester Action on Climate Change, Winchester, England, July 4, 2012, published by the Foundation for the Economics of Sustainability), accessed June 7, 2014, http://www.feasta.org/2012/07/13/climate-change-and-peak-oil-two-sides-of-the-same-coin/.

8. Richard Heinberg, *Snake Oil*, 22.

9. Richard Heinberg, *The End of Growth*, 48–152.

10. Brian Davey, "Peak Oil Revisited . . ." Foundation for the Economics of Sustainability, June 3, 2014, accessed June 7, 2014, http://www.feasta.org/2014/06/03/peak-oil-revisited/.

11. See for example the Mary Robinson Foundation, http://www.mrfcj.org/. Also see Climate Justice Taranaki, https://climatejusticetaranaki.wordpress.com/about/climate-justice/.

12. David Korowicz, "Tipping Point: Near-Term Systemic Implications of a Peak in Global Oil Production" (paper presented to The Foundation for the Economics of Sustainability and the Risk/Resilience Network, Dublin, Ireland, March 15, 2010), accessed June 7, 2014, http://www.feasta.org/documents/risk_resilience/Tipping_Point.pdf.

13. "Falling Oil Prices Could Lead Consumers to Spend More Later," *New Zealand Herald*, January 22, 2016, accessed January 23, 2016.

14. Arthur Berman, "IEA Offers No Hope."

15. "US Shale to OPEC: We Don't Care!" *CNBC*, December 4, 2014.

16. Hughes, Drill *Baby Drill*, ii–iii. SR Srocco, "CONDITION RED: Fracking Shale Is Destroying Oil & Gas Companies Balance Sheets," *SRSrocco Report*, August 13, 2014, 5, accessed August 27, 2014, https://srsroccoreport.com/condition-red-fracking-is-destroying-oil-gas-companies-balance-sheets/.

17. "About 50% of the Oil Produced in America in 2015 was provided by Hydraulic Fracturing, Reported the EIA," *Hydrogen Fuel News*, March 17, 2016, http://www.hydrogenfuelnews.com/tag/fracking-oil-production/.

18. "Unconventional Oil Revolution to Spread By 2019," International Energy Agency, Paris, June 17, 2014, accessed April 23, 2015, https://www.iea.org/newsroomandevents/pressreleases/2014/june/unconventional-oil-revolution-to-spread-beyond-north-america-by-end-of-decade.html.

19. "Hydraulic Fracturing: An Environmentally Responsible Technology for Ensuring Our Energy Future," Baker Hughes Corporation white paper, accessed March 3, 2015, http://www.bakerhughes.com/news-and-media/resources/white-papers/hydraulic-fracturing-white-paper.

20. Heinberg, *Snake Oil*, 37.

21. Engdahl, "America: The New Saudi Arabia?"

22. "Hydraulic Fracturing," Baker Hughes Corporation.

23. Giancomo Luciani, "The Promise and Perils of Fracking," *International Relations and Security Network*, January 9, 2014, accessed November 9, 2014, http://www.isn.ethz.ch/Digital-Library/Articles/Detail/?coguid=9c879a60-8a40-14e8-76c3-2c016ae9096c&lng=en&id=175268.

24. Hughes, *Drill Baby Drill*, 41.

25. See Hughes, *Drill Baby Drill*, ii.

26. Heinberg, *Snake Oil*, 30–31; also Srocco, "CONDITION RED."

27. Heinberg, *Snake Oil*, 30–31.

28. Engdahl, "New Saudi Arabia?"

29. Arthur Berman, "U.S. Shale Gas: Magical Thinking and the Denial of Uncertainty" (presentation at a workshop on Environmental and Social Implications of Hydraulic Fracturing and Gas Drilling in the United States: An Integrative Workshop for the Evaluation of the State of Science and Policy, Nicholas School of the Environment, Duke University, Durham, NC, January 9, 2012).

30. Kurt Cobb, "Orwellian Newspeak."

31. Engdahl, "The New Saudi Arabia?" Also Jeff Goodell, "The Big Fracking Bubble," 2.

32. Srocco, "CONDITION RED, 6."

33. IPCC, 2014: Climate Change 2014: Synthesis Report. Contribution of Working Groups I, II and III to the Fifth Assessment Report of the Intergovernmental Panel on Climate Change (Core Writing Team, R. K. Pachauri and L. A. Meyer [eds.]). IPCC, Geneva, Switzerland, 151 pp, accessed July 16, 2015, http://ar5-syr.ipcc.ch/.

34. Woodrow Clark, "The End of the Big Oil and Gas Game Has Come," *Huffington Post*, November 6, 2014.

35. Robert Howarth, Renee Santoro and Anthony Ingraffea, "Methane and the Greenhouse-Gas Footprint of Natural Gas from Shale Formations," *Climatic Change* 106 (2011): 679–690, accessed September 7, 2014, doi 10.1007/s10584-011-0061-5.

36. Mark Hertsgaard, "Scientists Warn: The Paris Climate Agreement Needs Massive Improvement," *The Nation*, December 11, 2015. See also Robert Howarth, "Methane and the Greenhouse Gas Footprint of Shale Gas" (presentation to the 100% Renewable Denton Town Hall Meeting, University of North Texas, Denton, Texas, March 25, 2016), accessed April 4, 2016, http://www.scribd.com/doc/306019583/Methane-and-the-Greenhouse-Gas-Footprint-of-Shale-Gas.

37. "U.S. Oil Fields Tagged as Culprit in Global Air Pollution Uptick," *The Philadelphia Tribune*, April 29, 2016, accessed May 5, 2016.

38. Chivers, "Fracking: the Gathering Storm," 11–12.

39. *Wikipedia*, "The Carbon Bubble," accessed November 15, 2015, https://en.wikipedia.org/wiki/Carbon_bubble.

40. Damian Carrington, "Carbon Bubble will Plunge the World into Another Financial Crisis – report," *The Guardian*, April 19, 2013, accessed April 23, 2013.

41. Arthur Neslen, "Climate Risks could Wreak Havoc on Financial Markets, EU Watchdog Warns," *The Guardian*, February 12, 2016, accessed February 13, 2016.

42. Peter Benson and Stuart Kirsch, "Capitalism and the Politics of Resignation," 464.

43. Jeffrey Jacquet, "Energy Boomtowns and Natural Gas: Implications for Marcellus Shale Local Governments and Rural Communities," Rural Development Paper 43, The Northeast Regional Center for Rural Development, Pennsylvania State University, 2009, accessed May 23, 2013, www.nercrd.psu.edu.

44. Jacquet, "Energy Boomtowns," 5. Also see Katheryn Brasier et al., "Residents' Perceptions of Community and Environmental Impacts from Development of

Natural Gas in the Marcellus Shale: A Comparison of Pennsylvania and New York Cases," *Journal of Rural Social Sciences* 26, 1 (2011): 32–61.

45. Susan Christopherson, "Marcellus Hydro-Fracturing: What Does it Mean for Economic Development?" (presentation, Department of City and Regional Planning, Cornell University, April 7, 2011), accessed September 2, 2013, from Catskill Mountainkeeper, www.catskillmountainkeeper.org/.

46. Janette Barth, "Critique of PPI Study on Shale Gas Job Creation," January 2, 2012, accessed February 28, 2013, Catskill Mountainkeeper, http://www.catskill citizens.org/barth/JMB_Critique_of_PPI_Jan_2_2012.pdf.

47. Christopherson, "Marcellus Hydro-Fracturing."

48. "The Economic Costs of Fracking," Catskill Mountainkeeper, 2011, accessed November 16, 2014, http://www.catskillmountainkeeper.org/our-programs/fracking/ whats-wrong-with-fracking-2/economics/.

49. "Economic Costs," Catskill Mountainkeeper, 51–53.

50. David Richardson and Richard Denniss, "Mining the Truth," 20.

51. Barth, "Critique of PPI Study." For similar findings see Food and Water Watch, "Exposing the Oil and Gas Industry's False Jobs Promise for Shale Gas," Washington, DC, 2011, accessed July 21, 2014, http://www.foodandwaterwatch.org/ insight/exposing-oil-and-gas-industrys-false-jobs-promise-shale-gas-development. Frank Mauro et al., "Exaggerating the Employment Impacts of Shale Drilling: How and Why," The Multi-State Research Collaborative, 2013, https://pennbpc.org/ sites/pennbpc.org/files/MSSRC-Employment-Impact-11-21-2013.pdf. Glen Greer, Michelle Marque and Caroline Saunders, "Petroleum Exploration and Extraction Study," Report to the Gisborne District Council. Agribusiness and Economics Research Unit, Lincoln University, 2013, accessed May 9, 2013, www.gdc.govt.nz/ assets/Files/.../EnvirolinkpetroleumimpactsstudyFINAL24Jan13.pdf

52. See Christopherson, "Marcellus Hydro-Fracturing"; also "Economic Costs," Catskill Mountainkeeper.

53. "The Economic Costs," Catskill Mountainkeeper.

54. Dutzik, Ridlington, and Rumpler, "The Costs of Fracking," 30.

55. Food and Water Watch, "Fracking Colorado: Illusory Benefits, Hidden Cost," Issue Brief, Denver, August 2013, accessed July 21, 2014, http://www.foodandwater watch.org/insight/fracking-colorado.

56. "Fracking in Australia: Gas Goes Boom," *The Economist,* June 2, 2012, accessed May 8, 2015.

57. Mark Broomfield, *Support to the Identification of Potential Risks for the Environment and Human Health Arising from Hydrocarbons Operations Involving Hydraulic Fracturing in Europe*, Report to the European Commission Directorate-General Environment, AEG Plc, ED57281, 17c. Didcot, UK, August 10, 2012, Table ES1, vi, accessed February 12, 2015, http://ec.europa.eu/environment/integration/ energy/pdf/fracking%20study.pdf.

58. United Nations World Water Development Report, Vol 1 *Water and Energy* (UNESCO, Paris, 2014), 23, http://unesdoc.unesco.org/images/0022/002257/225741E .pdf.

59. Kevin Zeese and Margaret Flowers, "US Climate Bomb is Ticking." Also Monika Freyman, "Hydraulic Fracturing and Water Stress: Water Demand by the Numbers," Ceres, 2014, 20ff, accessed June 1, 2015, http://www.ceres.org/resources/reports/hydraulic-fracturing-water-stress-water-demand-by-the-numbers.

60. Heather Cooley and Kristina Donnelly, *Hydraulic Fracturing and Water Resources: Separating the Frack from the Fiction*, Pacific Institute, Oakland, CA, 2012, 15, accessed February 12, 2014, http://pacinst.org/publication/hydraulic-fracturing-and-water-resources-separating-the-frack-from-the-fiction/.

61. See Karen Campbell and Matt Horne, "Shale Gas in British Columbia: Risks to BC's Water Resources," The Pembina Institute, Calgary, 2011, accessed May 29, 2013, https://www.pembina.org/reports/shale-and-water.pdf.

62. E.g., US Senator James Inhofe, R-Oklahoma quoted in *Popular Mechanics*, "Is Fracking Safe? The 10 Most Controversial Claims about Natural Gas Drilling," October 1, 2015, accessed November 21, 2015, http://www.popularmechanics.com/science/energy/g161/top-10-myths-about-natural-gas-drilling-6386593/. Nicholas Ortiz, Manager, Production Regions and Property Tax Issues for the Western States Petroleum Association of the Western States Petroleum Association, quoted in *KCET City News*, "L.A. Takes Big Step toward Fracking Moratorium," February 28, 2014. David Guglielmo, Halliburton Country Manager for Australia speaking at the 2013 New Zealand Petroleum Summit, quoted in the *Taranaki Daily News*, "Oil, Gas Sector Told to Improve Communication," October 2, 2013.

63. In 2014 the Pennsylvania Department of Environmental Protection reported 243 cases of drinking water contamination linked to shale gas production between 2008 and 2014. See also "Hydraulic Fracturing for Oil and Gas and Its Potential Impact on Drinking Water Resources," Environmental Protection Agency, June 2015, http://www.epa.gov/hfstudy/study-potential-impacts-hydraulic-fracturing-drinking-water-resources-progress-report-0.

64. Stephen Osborn et al., "Methane Contamination of Drinking Water Accompanying Gas-well Drilling and Hydraulic Fracturing," *Proceedings of the National Academy of Sciences* 108, 20 (May 17, 2011): 8172–8176.

65. Anthony Ingraffea et al., "Assessment and Risk Analysis of Casing and Cement Impairment in Oil and Gas Wells in Pennsylvania, 2000–2012," *Proceedings of the Academy of Sciences of the United States of America* 111, 30 (July 29, 2014): 10955–10960.

66. Richard Davies et al., "Oil and gas wells and their integrity: Implications for shale and unconventional resource exploitation," *Marine and Petroleum Geology*, 56 (September 2014): 239–254.

67. Campbell and Horne, "Shale Gas in British Columbia."

68. Dutzik, Ridlington, and Rumpler, "The Costs of Fracking," 14.

69. Campbell and Horne, "Shale Gas in British Columbia," 16.

70. "Drilling Boom Means More Harmful Waste Spills," *ABC News*, September 8, 2015, accessed September 16, 2015.

71. Personal communication November 18, 2015 from Ms Tina Ngata, Environmental Studies Tutor, Te Waananga o Aotearoa, Gisborne, New Zealand, see http://www.authorstream.com/Presentation/tina282217-2660219-three-myths/.

72. "Gas Drilling and Your Health," Catskill Mountainkeeper, 2013, accessed November 16, 2014, http://www.catskillmountainkeeper.org/our-programs/fracking/whats-wrong-with-fracking-2/health-impacts/.

73. Robert Jackson, "Fracking: What We Know and Don't Know About its Impacts on Water" (presentation to the Moos Family Speaker Series on Water Resources, College of Biological Sciences, University of Minnesota, January 30, 2014), accessed April 4, 2014, http://mediasite.uvs.umn.edu/Mediasite/Viewer/?peid=e65ac 4a5549e43deb76e1e1ef27a2a2a.

74. "Fracking and Water Consumption," SourceWatch, 2013, accessed March 11, 2013, http://www.sourcewatch.org/index.php/Fracking_and_water_consumption.

75. See Anne C. Epstein, "Health Risks of Living Near Natural Gas Development" (presentation to the 100% Renewable Denton Town Meeting, University of North Texas, Denton, Texas), accessed April 4, 2016, http://www.scribd.com/doc/306021913/Health-Risks-of-Living-Near-Natural-Gas-Development.

76. Seth Shonkoff, Jake Hays and Madelon Finkel, "Environmental Public Health Dimensions of Shale and Tight Gas Development," *Environmental Health Perspectives* 122, 8 (August 2014): 787–795.

77. Ian Urbina, "Drilling Down: Regulation Lax as Gas Wells' Tainted Water Hits Rivers," *New York Times,* February 26, 2011, accessed February 25, 2013.

78. Dutzik, Ridlington, & Rumpler, "The Costs of Fracking," 17–18.

79. Jackson, "Fracking: What We Know and Don't Know."

80. See Theo Colborn et al., "Natural Gas Operations from a Public Health Perspective," *Human and Ecological Risk Assessment: An International Journal* 17, 5 (September 20, 2011): 1039–1056.

81. Food and Water Watch, "Fracking Colorado," 3.

82. Donald Schweitzer, "The Myth of Purifying Fracking Water in Saudi America: The Competition between Food, Drink and Energy Needs," *Truthout,* January 29, 2013, accessed February 19, 2013, http://www.truth-out.org/opinion/item/14113-the-myth-of-purifying-fracking-water-in-saudi-america-the-competition-between-food-drink-and-energy.

83. William Ellsworth, "Injection-Induced Earthquakes," *Science* 341, 6142 (July 12, 2013), doi:10.1126/science.1225942.

84. Jackson, "Fracking: What We Know and Don't Know."

85. Nicholas van der Elst et al., "Enhanced Remote Earthquake Triggering at Fluid-Injection Sites in the Midwestern United States," *Science* 341, 6142 (July 12, 2013): 164–167, doi: 10.1126/science.1238948.

86. Natalie Starkey, "Pumping Water Underground could Trigger Major Earthquake, say Scientists," *The Guardian.* July 11, 2013, accessed March 16, 2014.

87. Starkey, "Pumping Water Underground."

88. William Freudenberg and Robert Wilkinson, eds, *Equity and the Environment*, Research in Social Problems and Public Policy 15 (Amsterdam, Elsevier, 2008): 3.

89. Heinberg, *Snake Oil*, 97–100.

90. Daniel Franks, David Brereton and Chris Moran, "Managing the Cumulative Impacts of Coal Mining on Regional Communities and Environments in Australia,"

Impact Assessment and Project Appraisal 28, 4 (December 2010): 299–312, doi: 10.3152/146155110X12838715793129.

91. Heinberg, *Snake Oil*, 99.

92. See M. Finer et al., "Oil and Gas Projects in the Western Amazon: Threats to Wilderness, Biodiversity and Indigenous Peoples," *Plos one* 3, 8 (2008): 1–9. John Altman and David Martin eds., *Power, Culture, Economy: Indigenous Australians and Mining*, Research Monograph 30, Centre for Aboriginal Economic Research, Australian National University, Canberra (2009). Paul Cleary, *Mine Field: The Dark Side of Australia's Resources Rush* (Collingwood, Victoria: Black Inc., 2012). G. Keucker, "Fighting for the Forests: Grassroots Resistance to Mining in Northern Ecuador," *Latin American Perspectives* 34, 2 (2007): 94–107.

93. Patricia Widener, "Community Responses to the Social, Economic and Environmental Impacts of Oil Disasters and Natural Resource Extraction" (presentation to a public seminar on issues around oil and gas development, Gisborne District Council chambers, Gisborne, New Zealand, October 16, 2013).

94. Sources: Brasier et al., "Residents' Perceptions." William Freudenburg, "Women and Men in an Energy Boomtown: Adjustment, Alienation, Adaptation," *Rural Sociology* 46, 2 (1981): 220–244. J. S. Gilmore, "Boom Towns May Hinder Energy Resource Development: Isolated Rural Communities Cannot Handle Sudden Industrialization and Growth without Help," *Science* 191 (1976): 535–540. Jacquet, "Energy Boomtowns." John McChesney, "Oil Boom Puts Strain On North Dakota Towns," *National Public Radio*, December 2, 2011.

95. New Zealand Parliamentary Commissioner for the Environment, *Evaluating the Environmental Impacts of Fracking in New Zealand: An Interim Report*, Wellington: New Zealand Government, November 2012. Also New Zealand Parliamentary Commissioner for the Environment, *Drilling for Oil and Gas in New Zealand: Environmental Oversight and Regulation*, Wellington: New Zealand Government, June 2014.

Chapter Three

New Zealand Government Efforts to Expedite Petroleum Development

Recent New Zealand governments have adopted the practice of translating their winning election platform into a strategic policy framework to guide departmental policy work, budget planning and legislative reform. "Reform" being a principled-sounding term for getting rid of legislation that obstructs the ruling party's political agenda. The National coalition government during its initial 2008–2011 term adopted a policy framework for economic development called "The Business Growth Agenda" which included major expansion of the oil and gas industry.

This chapter examines how, in order to clear the way for its agenda, the National government sought to enact an ideologically driven package of reforms intended to systemically dismantle the legislative regime of resource management and sustainable development (broadly understood) that had evolved through successive governments over more than a quarter century. The previous Labour government's economic, environmental, and local government policies in particular were considered barriers to business growth and natural resource exploitation that had to be dispensed with "in the national interest." The petroleum industry, along with industry-linked financial and legal consultants, helped develop National's energy policies, legislative reforms, and regulatory regime.

Legislative and regulatory reform was not the only way that the National Party facilitated extractive industry expansion. The government, again in collaboration with the oil and gas industry, embarked on a coordinated program of ministerial PR spin and cheerleading, targeted information dissemination, agenda-driven funding, marginalization of "radical" environmentalists, and community disempowerment to pave the way for oil and gas industry expansion.

When sectors like the petroleum industry work behind the scenes in co-operation with the state to manage legislative, policy-making, and public relations processes to further their mutual interests with the tacit knowledge (based on mounting research evidence) that current policies or commercial operations could have harmful effects or entail unacceptable risks *regardless of how stringent a government's regulatory regime*,[1] that constitutes collusion. In this chapter we will examine efforts by opposition parties, environmental and recreational organizations, and community activists to challenge National government policies, present countervailing evidence against these policies, and make it difficult for oil and gas companies to expand their operations. We will also note how this contention is reshaping long-standing relations between the government and community sector, a transformation we will explore in more depth in later chapters.

NEW ZEALAND POLITICS AND ENVIRONMENTALISM IN THE MODERN ERA

A Tradition of Popular Environmentalism

When the Second National government under Keith Holyoake was elected in 1960, they attempted to fast-track economic development by pursuing a policy called "Think Big." The state's role was to provide enabling legislation, subsidies, and tax payer funding to support major infrastructure projects like the Lake Manapouri hydro-electric scheme. Such a heavy-handed interventionist approach to revving up the economy was met with increasing nervousness within certain sectors of society. When a nascent coalition of environmentalists, scientists, recreational users, and farmers examined the provisions in the draft enabling legislation for raising the level of Lake Manapouri, they were shocked at the environmental implications. Within weeks the nation's first mass environmental movement, the Save Lake Manapouri Campaign, was underway. A three-year campaign managed to reverse some of the more draconian measures in the legislation.

Historian Michael King argues that "perhaps the most extraordinary feature of the [Save Manapouri] campaign was that, because it raised the fundamental question of how the nation should use its natural resources, it drew people from every possible political and ideological background."[2] In hindsight perhaps the rise of Kiwi "popular environmentalism" was not so surprising, given that similar movements were emerging in many industrialized countries following the publication in 1962 of Rachel Carson's *Silent Spring*. The endangered "environment" was becoming a concept people could relate to, reinforced by a nascent rediscovery by Pākehā (Europeans) of Māori understandings of Papatūānuku (Mother Earth) and kaitiakitanga (guardianship). For King the campaign's greatest legacy was "that it started a national debate on environmental issues, involving national and local body politicians, scientists, professional planners and members of the public. That debate . . . persisted long after the campaign itself had been won."[3] Both National and Labour parties were caught off guard by the mass upwelling of hitherto latent environmental activism. That same popular movement resurfaced in the 1980s when Robert Muldoon's National government attempted to push through the Clutha Dam project by overriding planning consent processes, and lost the 1984 election. The Manapouri and Clutha campaigns led to the passage of the Resource Management Act (RMA) in 1991, of which more will be said later.

One of the first things the Fifth Labour government under Helen Clark did after assuming power in 1999 was to develop a New Zealand Sustainable Development Strategy. The Strategy contained a series of measures to place sustainable development at the forefront of economic policy, departmental operations and community and local government activities.[4] Foremost among these were strengthening the RMA; passing the Local Government Act (2002) that required councils to adopt at a sustainable development approach to their activities, and initiating a Sustainable Development Programme of Action.

National's "Sustainable Growth" Approach

The debate over environmental and resource management has been reignited by the National government's economic growth policies and legislative reforms to promote petroleum industry expansion and commodity exports. The National Party under the leadership of John Key was elected in 2008 with support from the business and finance sector, urban middle-class professionals, and farming interests. National's framework for recovering from the global financial meltdown and boosting economic growth became known as the Business Growth Agenda. To achieve "sustainable" business growth, the government identified six key programs. One of these, aimed at accelerating mining and petroleum development, was paradoxically titled "Building Natural Resources." It was about putting in place measures to help local and international businesses exploit natural resources that would supposedly benefit the country. At the same time, the government promised to develop the "right systems and regulations" to ensure safe operations, and streamline resource management legislation.

National's 2009 Petroleum Action Plan

The petroleum industry has existed in New Zealand for over a hundred years, and is the country's fourth-largest export earner after dairy, meat, and wood products. Exploration and production activities currently generate around $300 million in company tax and over $400 million in royalties to the Crown,[5] even though petroleum companies pay only 12 percent of their profits in royalties which is low by world standards.

Early in National's first term, Minister for Economic Development Gerry Brownlee sought advice about the potential for the oil, gas, and mining sector to contribute to the economy. Share broking firm McDouall Stuart was contracted to prepare a report on steps the government could take to grow the sector. The report, called *Stepping Up,*[6] was written by a team headed by ex-oil man John Kidd, Head of Research at McDouall Stuart at the time. The report's main conclusion was blunt: "Weak political leadership, vision and strategy towards managing the national oil, gas, and mining estate [presumably referring to the previous Labour government] has been the key factor precluding the sector's development." Barriers to greater participation and investment by major international corporations included regulatory uncertainty, access to land and resources (e.g., under the Crown Minerals Act), and inadequate infrastructure. The report became the touchstone for National's legislative and regulatory agenda on resource extraction.

The report was followed in November 2009 by the release of the government's Petroleum Action Plan.[7] The Plan was developed in consultation with the industry and was composed of eight work streams comprised of government officials working in conjunction with industry representatives and consultants. It was a classic case of inter-sectoral collaboration between the government and the petroleum industry to implement a set of agreed strategies. The first work stream was about promotion: "Explicitly positioning Government as pro-active and pro-development of petroleum resources." The aim was to develop a "sustained communications strategy aimed at raising the profile of the petroleum sector and signaling government support for exploration and development activity." The target audiences for the campaign were international companies *and the New Zealand public*. Other work streams included a joint investment strategy for improving information (to E&P companies) about the country's petroleum resources; reviewing government's regulatory, royalty, and taxation arrangements for petroleum; and reviewing and amending legislation affecting the petroleum sector including offshore operations. The aim of the whole "collaborative" exercise, as industry legal consultant David Coull reflected several years later, was to progress "the continued responsible development of New Zealand's oil and gas resources."[8]

The New Zealand Energy Strategy

The Petroleum Strategy was incorporated into a more comprehensive New Zealand Energy Strategy 2011–2021.[9] The Strategy aimed to make the most of the country's "abundant energy potential through the *environmentally responsible* development and efficient use of the country's diverse energy resources."[10] Acting Energy Minister Hekia Parata reiterated in the document's introduction that expansion of the oil and gas industry could contribute to energy security and provide much-needed export earnings. The country had robust environmental regulations to ensure the environmental impacts of energy development and supply were "well managed." Anticipating environmental critics, she stated, "We cannot just turn off the tap in our journey to a lower carbon economy. As fuel costs continue to rise, a key challenge will be to reduce our reliance on petroleum while enabling New Zealanders to have access to competitively-priced energy."[11]

Opposition to National's "Sustainable Growth" Approach

Opposition to National's Business Growth Agenda came from a wide spectrum of political parties, academics, and environmental organizations as well as left-leaning economic commentators and new generation business people. The criticisms centered on accusations that (a) the government's export-led growth drive would have severe environmental consequences, (b) the "sustainable growth" approach was an oxymoron and contrary to authentic sustainable development,[12] and (c) the government was fiddling while Rome burned rather taking the leadership in developing a national strategy for transitioning to a clean-energy, high-technology economy.

For example, business commentator Rod Oram accused Prime Minister Key's government of systematically dismantling environmental programs and regulations including sweeping reforms to the RMA, the Local Government Act and EEZ regulations to remove environmental "barriers" to growth. The government's aims, Oram claimed, "could only be supported if there was proper pricing of externalities. These are the consequences of economic activity, such as pollution and climate change, experienced by unrelated third parties."[14] Pricing in externality costs wasn't something the government was prepared to include in its economic development policy-making.

At the 2013 Wellington *Valuing Nature* conference, Victoria University Professor of Public Policy Jonathan Boston called for an "ecological revolution" at all levels of New Zealand society and in policy-making.[14] There was insufficient appreciation of nature's contribution to people's livelihoods, health, security, and prosperity, he claimed, particularly among politicians and the business community. As an example, he slated the Key government's dishonest approach to economic development, on the one hand proclaiming the virtues of "clean, green New Zealand" while on the other implementing policies and legislative reforms like RMA amendments that were undermining that very image and endangering the country's environment. During the same period, several environmental groups and NGOs released statements and disseminated reports critical of National's "growth at all costs" approach to economic development. A non-profit research group called Pure Advantage, backed by a number of prominent new generation business leaders, released a study arguing the case for shifting to a national policy of "Green Growth." The report even received support from the Commissioner for the Environment. The National government rebuffed Pure Advantage's proposals. Instead the Minister for the Environment established a green growth working group of his own, most of whose recommendations were shelved after the working group reported back.

Other groups and energy experts criticized the government for not leading the way in developing a transition strategy to a sustainable, low-carbon economy. Even the Royal Society of New Zealand waded in, pointing out that under current international policy settings including New Zealand's, the world was on target for a 3.5°C rise by end of the century which would have serious economic, social, and environmental consequences. Per head of population, New Zealand's emissions were in the top ten globally.[15] A central theme in this lobbying was a call to open up the country's governance structure so that all sectors of society could be included in such planning. The Commissioner for the Environment called for an inter-sectoral "collaborative process" to develop a plan for weaning the country off dependency on fossil fuels and addressing climate change.[16] Meanwhile, several environmental, scientific, and business organizations including the Royal Society of New Zealand released reports on the feasibility of transitioning to a sustainable, low-carbon economy in order to address claims by government politicians and the oil industry that a rapid transition would be impractical and devastating to the economy.[17] A year after an Australia-New Zealand Climate Change and Business Conference in Auckland in 2015 called for establishment of a cross-sectoral forum to develop a national transition plan, the new Minister for Climate Change Paula Bennett announced she would establish a working group to provide advice to the government. Not exactly the bold, collaborative initiative that the PCE, environmentalists, iwi, the Sustainable Business Network, the Insurance Council, and Local Government New Zealand were looking for.

NATIONAL'S LEGISLATIVE REFORMS TO REMOVE IMPEDIMENTS TO EXPANDED NATURAL RESOURCE EXPLOITATION

The government paid only passing attention to such criticism, at least initially. National's focus was on growing the economy by promoting large and small business, agriculture, forestry, and natural resource extraction—in an "environmentally responsible way." Its support came largely from constituencies that operated within or benefited from these important economic sectors. It was these constituencies, through trade associations and sector organizations like Business New Zealand, Federated Farmers, and the Petroleum Exploration and Production Association of New Zealand (PEPANZ), who lobbied for cutting bureaucratic red tape and removing legislative "barriers" to business expansion and extractive industry development. The government encouraged such lobbying since it helped build popular support for its legislative reform agenda. Senior ministers were also aware of the need to bring the public along to avoid provoking the kind of backlash that had led to the Save Lake Manapouri Campaign. As National entered its second term in 2011, their legislative reform agenda concentrated on four pieces of legislation:

• The Resource Management Act 1991 (RMA);
• Exclusive Economic Zone and Continental Shelf (Environmental Effects) Act 2012 (EEZCSA);
• The Crown Minerals Act 1991 (CMA); and
• The Local Government Act 2002 (LGA).

The Resource Management Act 1991 (RMA)

The RMA is New Zealand's principal natural resource management and environmental protection statute. It covers the environmental effects of human activities on land and in New Zealand's territorial sea out to twelve nautical miles offshore. The purpose of the act is to promote the sustainable management of natural and physical resources in a manner which enables people and communities to provide for their social, economic, and cultural well-being and for their health and safety. The act also requires the potential of natural and physical resources (with the exception of minerals) be preserved to meet the foreseeable needs of future generations. Planning for, regulating, and monitoring the environmental effects of oil and gas exploration and extraction is primarily a responsibility of regional and local authorities.

The government claimed natural resource management reforms were necessary to increase the "effectiveness and efficiency" of the RMA and "reduce costs, delays and uncertainty to business."[18] Resource decisions were to be driven by "economic opportunity and society's needs" while still achieving "good environmental outcomes." This was a major shift in emphasis from the intent of the act. As the reform package rolled out, it was not clear (at least not to opposition parties and environmental activists) how the government was going to achieve the contradictory outcomes of accelerating natural resource exploitation while protecting the environment.

The government decided to carry out RMA reforms in two phases. Phase 1, which hopefully would be more palatable to the public, had to do with "efficiency": standardizing regional plans, reducing administrative procedures, and introducing a collaborative stakeholder consultation option as an alternative to costly, time-consuming litigation before the Environment Court. The Minister for the Environment embarked on a consultation exercise, inviting submissions and meeting with local authorities and key environmental and recreational organizations. Much was made of the hundreds of submissions received by the Parliamentary select committee. In reality, as we have seen, the government's Business Growth Agenda was already set, and submissions from environmental NGOs opposing the reforms as favoring development were paid scant attention. After receiving feed-back from an appointed Technical Advisory Group, the Minister pushed ahead. The RMA (Simplifying and Streamlining) Amendment Act became law at the end of 2009.

Environmental organizations, recreational groups, local anti-oil protesters, and Māori activists were even more anxious about the government's second phase reforms which the National government signaled would include changes to the basic principles of the act, requiring greater priority be given to economic development. The rallying cry went out that it was time to defend the RMA.

The Environment Minister launched the next phase of reform in 2011 by establishing a second technical advisory group (TAG 2), composed of people with planning, resource management and environment court experience. They were charged with reviewing the principles in sections 6 and 7 of the RMA and "other matters."[19] The principles were key to the way in which the act was given effect in planning and decision-making. TAG 2 reported back to the minister in February 2012,[20] recommending (as environmentalists and Māori activists feared) changes to the principles in section 6 to give equal weighting to economic development,[21] placing a stronger emphasis in section 7 on "timely, efficient and cost-effective resource management processes," and clarifying that the term "mitigation" as used in the act was not about *protecting* the environment and natural resources but "managing" them "sustainably."[22]

There was a flurry of criticism of TAG 2's report from environmental and recreational groups. Fish and Game New Zealand accused TAG 2 of going outside its terms of reference in an effort to support the government's economic growth agenda. Removing terms like "protection" and "preservation" and inserting economic development into Sections 6 and 7 was "an out-and-out attack on the environment, paving the way for rampant and unsustainable development."[23] New Zealand Greens' Eugenie Sage argued the TAG 2's proposals were "a major assault on the act and on sustainable management. . . . This government's agenda is to weaken the RMA to advance its 'dig it, drill it, mine it, irrigate it' agenda for resource exploitation."[24] The nationally renowned Environmental Defence Society (EDS) was so concerned about the TAG 2's report that they established their own TAG composed of environmental lawyers and resource planners to provide "independent" advice on the Minister's terms of reference. It was a relatively toothless gesture, but the EDS report circulated widely and garnered considerable support among like-minded groups.

The Minister obviously caught the drift from public submissions on the TAG 2 report that there was strong opposition to altering the act's basic principles. When the Resource Management Reform Bill 2012 was eventually introduced to Parliament on 5 December 2012, it focused on standardizing local planning and making resource consenting more "efficient." No changes were proposed to the principles because the Minister could not garner support from government's coalition partners. The "efficiency" amendments became law in 2013.

In March 2013 Environment Minister Amy Adams released a discussion paper entitled "Improving Our Resource Management System" once again proposing significant changes to sections 6 and 7 of the act. The Minister claimed the RMA in its current form was uncertain, cumbersome and too litigious[25]; businesses and developers wanted "certainty." She accused environmental groups of "scaremongering" and being "out of touch with New Zealanders." It was another attempt to caricature and marginalize environmental "extremists" from the rest of the population.

Unfortunately for the government, the ploy backfired. The Commissioner for the Environment, Te Arawa Iwi Coalition, Fish and Game New Zealand, and EDS were among thousands of organizations and individuals who made submissions on the discussion document.[26] EDS sought and disseminated legal advice arguing that the government's proposed changes to sections 6 and 7 would create a "shopping list" from which decision-makers could pick and choose, rather than the hierarchy of priorities currently set out in the act.[27] For example, protection of outstanding landscapes was treated as a more important consideration compared with "efficient use of resources" (e.g., oil and gas drilling). EDS issued a press release lambasting the minister for her public assurances that the proposed amendments would not erode environmental protections.[28]

Fish and Game New Zealand commissioned advice on the proposed reforms from Hon. Sir Geoffrey Palmer, QC, ex-Labour prime minister and the minister responsible for passage of the Resource Management Act in 1991. Palmer maintained that historically "the Resource Management Act represented a deliberate shift on the part of New Zealanders away from *economic advancement at any cost towards long-term economic and environmental sustainability.* It expressly acknowledged that the state of the natural environment and New Zealand's economic development were inextricably linked."[29] Based on several years of public consultation, the act signaled a national consensus to pursue a balanced approach to economic development and environmental sustainability. In Sir Geoffrey's view, "the Government's proposals fundamentally erode that commitment to sustainability" and would weaken "the environmental protection offered by the Act." He noted this was not the first attack on the act; environmental protections had been whittled away over time. In a damning summary Sir Geoffrey stated,

Under the rhetoric of efficiency key aspects of the RMA framework have been significantly eroded over time—for example, by increasing the ability of central government to intervene in decision making, reducing opportunities for public participation in decision making processes, and limiting the capacity for judicial supervision by the Environment Court. *The incremental effect has been to subtly tilt the framework of the RMA towards the facilitation of national development in order to meet Government economic policy. Thus, there has been a gradual creep away from sustainability* and back towards the centralised planning that the Act was originally introduced to replace.[30]

Sir Geoffrey's legal opinion piece circulated widely and became a touchstone for criticisms of government's proposed reforms by environmental and recreational groups as well as opposition parties. More than 14,000 submissions were received on the 2013 discussion document, 99 percent of which expressed opposition or criticized specific aspects of the proposed amendments.[31] The minister saw the writing on the wall. When the National Party's coalition partners the Māori Party and the United Future Party refused to support changes to sections 6 and 7, and with the 2014 election campaign looming, further work on the next RMA Amendment Bill was shelved.

With growing restiveness in the electorate about climate change and perceived threats to New Zealand's "clean, green image," top National party officials decided to remove RMA reform from the party's 2014 election platform. To head off further organized opposition, Prime Minister Key met with a deputation of environmental and recreational representatives and committed to seeking a "less confrontational" approach to RMA reform if National were re-elected."[32]

When election night was over, National had won a third term in office with an impressive 48 percent of the vote. Some media commentators predicted "further erosion of the Resource Management Act" and "continued support for mineral and oil exploration" as a consequence of National's victory. [33] Federated Farmers and Business New Zealand backed claims by some National politicians that the party had a clear mandate for further RMA reforms.

Environmental and recreational groups remained hopeful, particularly when Nick Smith was re-appointed as Minister for the Environment in a post-election Cabinet reshuffle. He had obviously been given the portfolio to kick-start RMA reforms. He was seen by some as the "most environmentally aware National minister" and had gained kudos for establishing the collaborative approach to freshwater reform via the Land and Water Forum.[34] Behind the scenes, Smith (who was also Minister of Building and Housing) was contemplating a bold maneuver to refocus the RMA debate and hopefully garner wide popular support for more radical changes.

On January 21, 2015, Dr Smith made his move. Preparing to deliver his annual address to the Nelson Rotary Club, the Minister took the stage in front of several huge stacks of local authority resource management plans representing bureaucratic red tape and effectively declared war on section 6 and 7 of the RMA. It was great political theater. This time, however, the minister had changed the battle plan and caught almost everybody off guard. The minister's message was as simple as his misdirection strategy was clever. Everyone agreed that affordable housing, particularly in Auckland, had become an urgent issue. Further reform of the RMA was critical to addressing the housing crisis. That's why he was announcing the most significant overhaul of the RMA since its inception in 1991. The minister warned "tinkering with the Act won't do." Development had to be incorporated more specifically into the act to achieve "balance," albeit in a way the avoided "*unnecessary harm to the environment*."[35]

Off the public stage in coming months, Smith continued to talk about the need for a collaborative approach to legislative change. He was after all a member of the self-styled Blue-Greens within the National Party and had to contend with the more conservative wing who constantly pressed for further legislative reform to stimulate business growth. Most media commentators, opposition party spokespersons, and the public bought into Smith's maneuver and joined the debate over how much additional RMA reforms would actually improve the housing supply the situation. Few questioned (at least initially) whether the RMA was about urban development or whether it was actually about sustainable resource management and environmental protection. It was a classic magician's sleight-of-hand trick.

Smith's Nelson speech immediately raised concerns among environmentalists, recreational organizations and anti-fracking groups. A number of environmental groups convened at the annual ECO conference[36] in Taranaki in March 2015 to discuss a strategy for "defending the RMA." But their plans were soon overtaken by political developments in Northland.

On January 30, 2015, Northland National MP Mike Sabin resigned, sparking a by-election. Polls shortly before the March 28, election day showed New Zealand First MP Winston Peters, whose party was in coalition with National but who refused to back changes to RMA principles, had a commanding lead. Prime Minister Key, Nick Smith, and other MPs paid urgent last-minute visits to the region, warning voters that if Peters was elected government's planned overhaul of the RMA would be at risk because National would lose its parliamentary majority.[37]

In spite of National's efforts, Peters was elected with an overwhelming majority. National's campaign manager Minister of Economic Development Steven Joyce acknowledged Peters's election would make things more difficult regarding RMA reform. The government wouldn't be able to "advance the same level of reforms we would have," which Joyce said was a loss for Northland. Prime Minister Key was clearly peeved: "With the RMA, there's just no question that you've got to rip up what we've got now, go back to the drawing board and have another go."[38] Yet it was hardly the embarrassing back down some critics claimed. Negotiations continued behind the scenes with National's coalition partners to try to gain support for amendments of the act's principles, but it proved a difficult balancing act. The Māori Party was offered a bigger role for Māori in the Resource Management consenting process. That sparked criticism from Winston Peters, ACT leader David Seymour, and conservative National backbenchers. Eventually in late 2015 Environment Minister Nick Smith introduced the Resource Legislation Amendment Bill which critics warned would give the minister additional powers to intervene in local council plan-making and consenting processes, and further curtail the rights of citizens to be heard at consent hearings.

On November 1, 2015, the government surprised pundits and RMA defenders alike by announcing another reform initiative, this time aimed at rethinking the *entire urban and environmental planning regime* including the RMA. Announcing the initiative, Finance Minister Bill English declared the "whole country" would be involved in the rethinking process (clearly hoping to legitimize sweeping changes through widespread consultation as Labour had done in passing the RMA in 1991). The main instrument was to be a "blue skies" review by the Productivity Commission, a business and development-leaning body established by the government after the 2008 election. The process would take a year and was intended to be a "first principles" review looking "beyond the current resource management and planning paradigm and legislative arrangements to consider *fundamentally alternative ways of delivering improved urban planning and, subsequently, development*."[39] It would encompass resource management, environmental protection, conservation and local authority powers within its scope.

The terms of reference also had serious implications for local councils and communities. The review was to:

> Consider ways to ensure that the regime is responsive to changing demands in the future, how national priorities and the potential for new entrants [like the petroleum industry] can be considered alongside existing local priorities and what different arrangements, if any, might need to be put in place for areas of the country seeing economic contraction rather than growth.[40]

The Productivity Commission review was the beginning of a major multi-year push by the government to dismantle the RMA and put in place an entirely new urban planning, land use and resource management regime.[41] Now this was real sweeping reform!

The Crown Minerals Act 1991 (CMA)

This act establishes that the Crown owns all minerals under the ground.[42] It sets out how the Crown allocates rights to explore, prospect, or mine Crown-owned minerals including petroleum. The Minister of Energy and Resources allocates prospecting, exploration, or mining permits for a specific area which in turn are administered by New Zealand Petroleum and Minerals (NZPaM), an agency within the Ministry of Business, Innovation and Employment (MBIE). The act was amended in 1997, placing thousands of hectares of conservation land under a new schedule 4 which prohibited mineral exploration, mining, and other industrial activities.

After National won the 2008 election, and Minister of Energy and Resources Gerry Brownlee in conjunction with the Minister for Conservation Kate Wilkinson carried out a stock take of public conservation land with mining potential. In 2010 the ministers jointly released a discussion document called *Maximising Our Mineral Potential,* proposing relaxation of access requirements on schedule 4 land and removal of over 7,000 hectares of land from schedule 4 designation. Brownlee acknowledged in his media release that mining was an "emotive issue" and accused the "environmental lobby" of exaggerating the potential impacts. He maintained mining could contribute to the country's prosperity and called for a "rational conversation."[43]

Public reaction to the proposals was immediate and strident on both sides of the debate. Business New Zealand's Phil O'Reilly claimed the stock take showed the government was "acting judiciously on behalf of all New Zealanders."[44] Forest and Bird denounced the Brownlee-Wilkinson measures as a sham. Greenpeace reminded the public of the Think Big era that led to the Save Manapouri campaign. Within a few weeks, there was a groundswell of popular opposition. Over 30,000 people marched in protest down Auckland's Queen Street on May Day 2010. The Ministry of Economic Development received more than 37,000 written submissions on the discussion paper, most opposed to exploration and more mining on conservation land. The Parliamentary Commissioner for the Environment lodged a strongly worded submission suggesting the government had gotten the cart before the horse in opening up pristine land before assessing the true costs. Two months later, Ministers Brownlee and Wilkinson announced the government had listened to the public and was shelving its proposals.

However, Minister Brownlee wasn't quite ready to admit defeat. After further lobbying from the mining and petroleum industries helped bolster the government's case, on September 2012 he introduced the Crown Minerals (Permitting and Crown Land) Amendment Bill to Parliament. The bill established a two-tiered system to streamline permit management and make it easier for small-scale operations to proceed. It revised the existing purpose statement, adding the aim of promoting mining and providing for "efficient allocation" of resources (shades of Nick Smith's RMA reforms).

Simon Bridges assumed the Energy and Resources portfolio from Brownlee after a Cabinet reshuffle in January 2013. On March 25, 2013, shortly after the Crown Minerals (Permitting and Crown Land) Amendment Bill passed its second reading, Bridges brought a paper to Cabinet proposing urgent changes to the bill before it went to a final vote. Cabinet accepted the changes and agreed the bill should proceed under urgency without going back to a select committee for consideration. The public weren't to get a chance to make submissions on the changes.

Minister Bridges chose the long Easter holiday weekend, when media coverage would be limited, to announce the changes. The revisions to the bill, introduced to Parliament the following Wednesday, were aimed at preventing further protest disruptions to offshore exploration and drilling. Māori groups and Greenpeace had recently undertaken seaborne protests, in the former case leading Brazilian company Petrobras to cancel its exploratory operations in the Bay of Plenty.[45] Other protests were rumored to be immanent. The amended legislation created five new offences for interfering with mining and drilling operations at sea within a 500-meter exclusion zone. The new offences and police powers were portrayed by the minister as steps by a prudent and responsible government to ensure companies could go about their lawful business while maintaining the public's right of protest . . . at a distance.

Opposition parties and environmental groups were furious at the haste with which the bill was being rushed through Parliament. In a joint statement, a group of eminent New Zealanders, unions, human rights groups, and environmental organizations claimed the bill was "a sledgehammer designed to attack peaceful protest" and was "being bundled through Parliament without proper scrutiny despite its significant constitutional, democratic and human rights implications." More than 45,000 people eventually added their names to the statement.

It turned out there was more to the urgent changes to the bill than met the eye. It emerged that on February 14, two weeks before going to Cabinet with his paper, Bridges had met with representatives of the petroleum industry including Shell New Zealand.[46] Initially the minister denied he had received requests from industry representatives for the amendments to the bill. But when further inquiries confirmed the meeting with industry representatives had in fact taken place, Labour (who were not against deep-sea drilling though they were cautious about fracking)[47] accused the minister of misleading Parliament. The minister claimed the timing of the meeting was coincidental and the anti-protest amendments were never discussed.[48]

Documents subsequently released to Greenpeace under the Official Information Act revealed that six months earlier on September 4, 2012, Economic Development Minister Steven Joyce had met with Shell New Zealand's chairman Rob Jager (also board chairman of PEPANZ), business advisor Chris Kilby, and PEPANZ CEO David Robinson. Jager tabled a paper expressing industry concerns there was "insufficient legal authority" to prevent offshore protests, and that the government "has no teeth beyond 12 nautical miles to protect legitimate commercial activity." The documents also revealed Bridges met with Jager, Kilby, and Robinson (the same people who met with Steven Joyce) on February 14, 2013, just two weeks before he presented his a paper to Cabinet proposing urgent amendments to the bill. Bridges's paper noted that "the upstream oil and gas industry has sought a more robust government response to threats of, and actual, direct protest action."[49] It was obviously a reference to Jager's earlier paper to Joyce and his own meetings with industry officials. When pressed by journalists and environmental groups for details about the February 14 meeting, Bridges stated that "notes weren't taken" and he had nothing further to add. Suspecting that he had *directed* officials not to take notes (since the convention is that officials always take notes), critics and opposition parties accused the minister of a cover-up. The meetings and anti-protest legislation had all the hallmarks of collusion, but in the end there was only circumstantial evidence to support critics' claims. The Crown Minerals Amendment Act 2013 became law on May 19, 2013.

Exclusive Economic Zone Act and Continental Shelf (Environmental Effects) Act 2012 (EEZ Act)

The EEZ Act came into force in 2013 and established an environmental management regime for New Zealand's Exclusive Economic Zone and con-tinental shelf out to 200 nautical miles. The act classified regulated activities within the zone as either permitted, discretionary, or prohibited. Exploratory drilling and mining was initially classified as a discretionary activity. Com-panies wishing to undertake these activities were required to be granted a marine consent from the newly established Environmental Protection Agency (EPA). The application would then be publically notified, assessed, and a public hearing held before the consent was either granted or rejected.[50]

The petroleum industry preferred offshore exploration and drilling to be a permitted activity. PEPANZ had previously floated the idea (unsuccessfully) that all government mining permits should include resource consents. If off-shore oil and gas operations couldn't be classified as a permitted activity, then the industry lobbied for consent applications to at least be exempted from public notification. After all, that was how municipal authorities were handling most onshore resource consent applications. In December 2013, Environment Minister Amy Adams proposed amendments to the year-old EEZ Act creating a new activity classification: "non-notified discretionary." Operators would still have to apply for a marine consent, but there would be no public notification and no hearings.[51] Environmental groups branded the proposed regulations "The Anadarko Amendment" because it favored transnational E&P companies who already held permits to drill off New Zea-land's coast. The Environmental Defence Society's executive director Gary Taylor complained that "the international petroleum industry has lobbied in-tensively" to obtain the special concession prohibiting public involvement in the consent process.[52] He warned there was a high risk of regulatory capture. Greenpeace echoed Taylor's concerns about lack of public involvement in the consent process.[53] Ignoring criticisms from environmental groups and iwi, the minister announced in February 2014 the new "non-notified discretionary" classification would be included in amended EEZ Act regulations.[54] It was a "pragmatic option," she said, that would save the industry time and extra cost while providing "effective oversight and environmental safeguards." Without public involvement, it was another example of the government's "trust us, we know what we're doing" approach to resource management and environmental protection. It was also one of the government's many maneuvers to pave the way for expanded oil and gas exploration and drilling.

The battle was joined again in 2016 when the Marine Reserves Act came up for review. The National Party proposed in the 2014 election campaign to introduce a new Marine Protected Areas Act that would provide for more marine reserves in the country's EEZ. But a discussion document on the proposed act indicated its provisions would only apply within the twelve-mile territorial limit rather than out to the 200-mile nautical limit. Environmental critics once again claimed the government had caved in to lobbying by extractive industries and pressure from pro-development interests.[55] Submissions on the draft legislation reflected a national poll showing 76 percent of the public favored extending marine protections to the 200-mile limit. Oil and gas lobby group PEPANZ came out against extending the protections beyond twelve miles.[56]

The Local Government Act 2002 (LGA)

Local Government Act reforms were fundamental to National's agenda of systematically dismantling key aspects of the existing resource management/sustainable development regime in order to boost business growth and expand extractive industry.[57] The LGA 2002 was about strengthening "participatory" local government and pursuing outcomes-driven sustainable development. The act required local authorities "to play a broad role in promoting the social, economic, environmental, and cultural well-being of their communities [often referred to as "the four well-beings"], taking a sustainable development approach."[58] It also introduced a planning procedure called the "community outcomes process." This was intended to be a community-driven exercise facilitated by local authorities that identified local well-being outcomes, balanced among competing priorities, and suggested how different sectors of the community could help to achieve mutually-agreed outcomes. The result of the process was a Long-Term Council Community Plan.

During the 2008 election, the conservative ACT Party's leader, Rodney Hide, called for measures to reduce inefficiencies and costs of local government. When his party went into coalition with National following the election, Hide was given the Local Government portfolio. He immediately embarked on a review of the LGA 2002 and soon released a discussion paper titled *Smarter Government, Stronger Communities: Towards Better Local Governance and Public Services.*[59] From his public statements it was clear Hide had little time for sustainable development, nor it appears strong communities. He abhorred prolonged community consultation. He certainly didn't think it was the business of local councils to be involved in social services and community development.[60] Hide's LGA 2002 Amendment Act 2010 did away with the community outcomes process in the name of efficiency, but he deferred action on the purpose statement regarding sustainable development because of mounting opposition from community and environmental groups.

Following the 2011 election, National MP David Carter took over as Minister for Local Government. In March 2012, he announced another round of reforms called *Better Local Government*. Carter confirmed that making changes to the act was "part of the government's broader agenda relating particularly to its strategic priorities of building a more productive and competitive economy."[61] In June 2012 the Minister introduced LGA 2002 Amendment Bill (No. 2) into the House and, after muted feedback from local councils but loud objections from community and environmental groups, the act became law in December. The most crucial change, from the perspective of government's agenda of dismantling "barriers" to "sustainable growth," was a complete revision of the purpose statement. The new act deleted all references to the four well-beings and altered Section 3 (d) which explicitly directed councils to adopt a sustainable development approach in their activities. Instead, the amended act required local authorities "to play a broad role in meeting the current and future needs of their communities for good-quality local infrastructure, local public services, and performance of regulatory functions." In the words of one opposition party wag, the purpose of local authorities was now about "Rats, Roads and Rates" (taxes).

The second phase of local government reforms was about cutting red tape for businesses and developers, and reducing the burden of rates on property owners. In policy terms, the impact of the proposed changes was much wider. The minister appointed a Task Force that reported back in December 2012 with its recommendations. The Task Force noted at the beginning of their report that they had elected to broaden the definition of "efficiency" to encompass "effectiveness," including reference to effective community engagement.[62] Unfortunately, the group's final recommendations had the effect of reducing community engagement. Adopting the government's neoliberal perspective, the Task Force argued that the model of local government contained in Section 6 of the LGA 2002 had been "misconstrued" to mean "participatory democracy." Unlike "representative democracy" upon which the act was supposedly founded, the Task Force maintained participatory democracy was *inherently impractical and inefficient*.[63] Taken to the extreme, it meant mob rule.

Partly in response to the Task Force's report and the amendment bill before Parliament, a resolution was passed at Local Government New Zealand's

2013 conference calling on local government in New Zealand to be given constitutional protection. Delegates may have been stirred to action by a key-note address on the rise of "localism" in Western democracies by Oliver Hart-wich, Executive Director of the New Zealand Initiative. Hartwich maintained that localism could help preserve a free and democratic society and deliver government services more effectively based on sound economic principles.[64] However, he warned, powerful private and political interests "will always find reasons for centralising power and decision-making when it serves their own agenda."[65] In contrast to the UK government's support for localism, the National government moved in the opposite direction with the passage of the Local Government Act 2002 Amendment Act 2014. One of the crucial shifts in emphasis as far as regional planning was concerned was that central government had more influence over important development decisions, such as whether to support oil and gas development, since they were judged to be "in the national interest."

GOVERNMENT MANEUVERS TO PROMOTE PETROLEUM DEVELOPMENT

Early in its first term, National determined that expansion of the petroleum industry had the potential to make a significant contribution to export growth, could be done safely, and would not conflict with the country's efforts to address climate change. From the government's perspective, expansion of the petroleum industry was not a matter for debate (not a real one, at least) but of facilitation and public promotion. One is reminded of Sir Humphrey Appleby's admonition to the PM's secretary in the British comedy *Yes, Prime Minister*: "For goodness sake, Bernard, you don't open a national debate about an issue until the Government has made up its mind!!"[66] To help grow the petroleum industry, National adopted a number of what became standard manoeuvers and tactics.

"Factual" Information Dissemination

Over three successive terms, the National government became increasingly sophisticated at marketing the country's deep-sea and unconventional on-shore prospectivity and promoting the oil and gas industry to the New Zealand public. Nevertheless, public concerns remained over the environmental and health risks of fracking and the potential for catastrophic offshore oil spills. To address these concerns, the government made extensive use of "official" information dissemination and PR spin to underscore the industry's importance to the national economy. Such information was typically labeled "factual" (or alternatively "scientific," "evidence-based," or "objective") and distributed in a calculated manner to validate government's petroleum industry policies and programs, influence public perceptions, provide supporters with ammunition, and counter the supposedly emotive, ill-informed, and unscientific claims of critics. It also helped bolster the authority of the relevant official or minister releasing the information.

The primary conduit for government's "factual" information about the petroleum industry is MBIE and its subsidiary NZPaM. The ministry and its lead minister, Steven Joyce, take public relations and agenda-driven information dissemination very seriously. In May 2014 a report by the State Services Commission revealed that MBIE had accumulated a communications team of fifty-six people. Twenty of the fifty-six held traditional communications roles, while the remainder were involved in website development, design work, and marketing.[67] As concerns mounted over National's plans to expand deep-sea exploration and land-based unconventional oil and gas development, the Commissioner for the Environment initiated a two-stage study of fracking in 2012. At the time, environmental and anti-fracking groups were organizing protests and receiving considerable press coverage. Economic Development Minister Steven Joyce decided it was time to launch a campaign to set the record straight. What follows are a few examples of purposeful information dissemination from MBIE in support of the petroleum sector. They are instructive because they demonstrate the techniques used to manipulate supposedly authoritative "official" information for the government's own purposes.

One of MBIE's first salvos was a paper titled the "Economic Contribution and Potential of New Zealand's Oil and Gas Industry." Released in August 2012, the paper was produced by MBIE's Economic Development Group and based on studies commissioned from two economic consultancies.[68] To be sure the government's key messages got out, the project team's wordsmithing tactics included:

1. *Recontextualizing or exaggerating facts*: The writing team cited 2009 research by GNS Science claiming there was a "huge potential to discover new oil and gas reserves" because New Zealand was "scarcely explored." [69] A handful of similar studies have been produced, leading to extraordinary estimates by ministers and industry executives of the country's oil and gas reserves. Reserve estimates are typically provided by the petroleum industry itself.

2. *Dubious truth claims*: MBIE's paper asserted that oil and gas would continue to play an important role in the country's "energy mix" for the long-term future. The main reason to expand oil and gas development was

that it would support regional development and improve "our standard of living."[70]

3. *Trust us requests*: The paper contained standard assurances the government knew what they were doing, asking the reader to suspend critical judgment. "New Zealand has rigorous regulations around oil and gas exploration and production activities in order to manage the risks to health, safety and the environment."[71]

4. *Disguised guesswork*: The paper discussed the potential growth of the industry based on several scenarios using data from respected economic consultants. Closer examination revealed the scenarios were not only speculative but based on a number of highly questionable assumptions.[72]

A second government information report was published six months later, targeted at the country's East Coast which both the government and the petroleum industry hoped would be where the next boom occurred. Titled the "East Coast Oil and Gas Development Study,"[73] it was produced by another MBIE study team once again using commissioned technical papers. The study was legitimated by a working group convened at the behest of Economic Development Minister Joyce and composed of representatives of East Coast local councils and Business Hawke's Bay. Although MBIE's study began months before the working group was established, the final report was portrayed as a "collaboration" between the Ministry and East Coast local authorities. It was intended to "support informed dialogue between councils, communities and iwi about the potential benefits, impacts and risks of petroleum development across the East Coast, if such a development were to eventuate."[74] The report was used to support government's subsequent PR campaign to sell East Coast residents on the petroleum industry as we'll see later.

A third major piece of "factual" information was the "Petroleum and Minerals Sector Report," released jointly by Economic Development Minister Steven Joyce and Energy and Resources Minister Simon Bridges in September 2013. The report was published during a period of intense parliamentary debate and media coverage over National's plans to amend the purpose and principles of the RMA. Releasing the report, Minister Joyce repeated a standard government and industry mantra: "If we want more and better paying jobs and more money to invest in our schools and hospitals then we need to keep making the most of our abundant energy and minerals potential through environmentally responsible development."[75] He noted the report was one of a series on different sectors of the economy "designed to support the government's Business Growth Agenda."

Six months later, Joyce announced MBIE would be producing a series of regional potential studies to inform (and no doubt guide) local thinking about economic development. Not surprisingly, the first study was targeted at the East Coast[76] including a section on the contribution oil and gas development could make to the region, repeating the speculative scenarios from MBIE's earlier East Coast oil and gas study.

In 2015 after several of these studies had been churned out, Minister Joyce launched a second phase of government's regional development policy. Again beginning with the East Coast, Joyce announced his officials would be working with Activate Tairāwhiti, Gisborne region's economic development agency and business leaders to formulate what he later claimed was a "locally developed" Action Plan for regional growth. Joyce told a Gisborne Chamber of Commerce meeting he wanted regional development planning to be driven by the business sector because in past regional planning, local government had "run off" with the process.[77] Oil and gas was expected to figure large in the Action Plan. Meanwhile, in Northland MBIE officials had been working with Northland, Inc., the Northland Regional Council's economic development arm, as well as iwi and business leaders to prepare a Northland Regional Development Action Plan. Once again the process was restricted to 'key stakeholders' and excluded the general public.

NZPaM is another important conduit for government's "factual" petroleum and minerals information and marketing. It was established in 2011 after restructuring of the Crown Minerals agency. The aim was to strengthen the government's capability to market the potential of the country's oil and gas basins, although the agency's regulatory and promotional functions are now separate.[78] NZPaM's basic role is to administer the Crown Minerals Act. That includes consulting with local authorities and iwi about proposed mining permits, and providing information to landowners and communities about extractive industries.

Sometimes with the best intentions, "factual" dissemination maneuvers go wrong and result in big headaches for an agency and its minister. In 2013 NZ-PaM decided to organize a series of "public information" road shows about oil and gas development. It was shortly after the Parliamentary Commissioner for the Environment had released her 2012 interim report on fracking and during consultation on the latest round of petroleum exploration block offers. Aware of growing public controversy, NZPaM sent out staff for confidential discussions with various local councils about the purpose and procedures for what were initially conceived of as a series of invitation-only consultations.

Kaikoura District appears to have been the agency's first attempt at organizing an invitation-only information meeting to discuss the recently announced block offer covering the district. The area was a popular tourist destination for whale-watching. The proposed exploration block included deep-sea exploration covering a tourist whale-watching area so community interest was keen. Complying with the letter of the Crown Minerals Act, NZPaM arranged to meet privately with Kaikoura District councilors and representatives of local iwi (tribe) Te Korowai o Te Tai o Marokura in late October 2013. Unfortunately for the organizers, word leaked out and members of the general public and media clamored to attend. Shortly before the day of the meeting, an item appeared in the local newspaper saying the meeting had been canceled.[79] Instead, two open meetings were scheduled for a fortnight later at a local marae (communal gathering place). NZPaM, presumably anticipating heated discussion, revised the meeting format to try to maintain a semblance of control. Instead of discussing the proposed exploration block that would cover the district, a panel of government bureaucrats were invited to explain the government's "world class regulatory regime." Local Māori leader Sir Mark Solomon was asked to play the role of neutral facilitator. NZPaM began with a promotional slideshow about the importance of the oil and gas industry to New Zealand. After presentations from the other officials, questions were invited from the floor. Discussion at both meetings quickly became heated. A journalist attending the hui (meeting) later reported "the information did little to allay community fears of an environmental accident in the region." Officials were asked to take back the message that "the community didn't want to see deep-sea drilling off the coast of Kaikoura or elsewhere in New Zealand."[80]

By the time of a Dunedin road show early the following year, NZPaM had revised the format further, this time to cover just the government's regulatory regime. The underlying purpose was still to convince the public that the oil and gas industry was well regulated and safe. The Dunedin hui began with a welcome, introductions, and much abbreviated presentations about complex regulatory information. The audience was restricted to asking questions of clarification but in spite of the moderator's efforts, a lively and heated discussion broke out about the risks of deep-sea exploration and fracking.

NZPaM officials evidently decided further format changes were needed to manage the "discussion." The next opportunity came a month later in Northland. The agency had already experienced difficulties with Māori protesters in the region. Staff had been thrown off one marae and shouted down at another.[81] This time, after confidential discussions with the Northland Regional Council (NRC) and representatives of Norwegian company Statoil which had a permit to explore off the Northland coast, a rather clumsy solution was hit upon to control the debate and contain protest. The regional council agreed to do the front running with the media. The plan was to hold two "workshops," the first of which would be invitation-only for regional councilors and the council's Māori Advisory Committee. NRC chairman Bill Shepherd told media this was so elected representatives could hear information from regulatory officials and Statoil, and "have questions answered in a *structured* manner."[82] Each workshop was divided into two half days. The morning session was NZPaM's standard roadshow; the afternoon involved presentations and a Q&A session with Statoil representatives. Shepherd promised a second workshop "at a date to be confirmed . . . for those who oppose exploration to present information and answer questions."[83] Although the closed workshop was scheduled three days after Sheperd's announcement, Greenpeace and the Northland Stop Statoil movement got wind of the planned workshop and organized protests outside the venue.

In reality NZPaM's "roadshows" were more of an exercise in mis-information and citizen disempowerment, given mounting evidence of adverse effects of unconventional and deep sea drilling, and growing public concerns about climate change being raised at these meetings.

Official Propaganda, Cheerleading, and Jawboning

Propaganda can be a subtle form of state coercion as well as simply influencing public opinion. Politicians of course would deny any of their rhetoric is propaganda, preferring to talk about "getting the messages out" about policies like oil and gas development "in the national interest." Officially sanctioned "messages" (propaganda), ministerial cheerleading, and jawboning are aimed at shaping public perceptions, fabricating, and managing the "debate," and (to borrow Herman and Chomsky's poignant phrase) "manufacturing consent."[84] In New Zealand as no doubt elsewhere such activities are frequently carried out in cooperation with the petroleum industry whose input is sought in preparing policy, developing departmental PR campaigns, and compiling authoritative-seeming reports to disseminate to the public.

Government propaganda and ministerial cheerleading has tended to reach a crescendo during the annual block auction process and the Advantage New Zealand Petroleum Summit. Minister of Energy and Resources Simon Bridges has a reputation as government's leading cheerleader for the oil and gas industry.[85] Paradoxically, he is also responsible for regulating the industry for whom he so enthusiastically advocates. It is instructive to examine some of Bridges's propagandizing efforts during the time of the 2013 and 2014 block offer auctions and related oil and gas conferences.

Bridges proved a fast learner in the craft of marketing and public indoctrination. In April 2013 he appeared at the annual Petroleum Summit to announce the government's new exploration block offer. Speaking to an audience that included a number of international guests, he enthused about what the industry and government could "achieve together."[87] His presentation emphasized four themes from what was rapidly becoming his standard promotional repertoire:

- Opportunity: "In oil and gas terms, New Zealand truly is a land of opportunity. . . . We are recognised as one of the world's most promising regions, but we remain relatively unexplored."
- A world-class regulatory environment providing business certainty: "We have a strong regulatory environment that is supportive of responsible exploration and development."
- Legislative reform for greater efficient and effective resource management: changes to the Crown Minerals Act "will give New Zealanders confidence and certainty that regulators have the tools to adequately scrutinise all aspects of exploration and production."
- Safe, environmentally responsible development: The government is committed to "ensuring New Zealand has a world-class regulatory system that ensures the safety of its people and its environment, alongside greater resource development."[87]

Similar themes were highlighted in a discourse analysis of mainstream print media stories on oil and gas development (particularly deep-sea exploration and production) by Diprose, Thomas, and Bond.[88] Claims that such development was essential for the country's economic development and/or that the state's regulatory regime insured industry operations were safe and environmentally responsible featured in 42 percent and 33 percent of the articles, respectively.

At the April 2014 Advantage New Zealand conference, with an election looming and environmental critics to answer, Bridges decided to begin with a nod to climate change. Adopting recent industry spin, he stated:

> We [oil people and politicians] want our citizens and countries to continue to develop and prosper. We need to reduce carbon emissions from our energy use and respond to climate change. . . . Oil and gas will continue to play a key role in ensuring our energy supply is reliable and affordable.[89]

The country needed to "reduce carbon emissions from our energy use and respond to climate change," he said. But the government's policy was neither exclusively renewable nor non-renewable; "fossil fuels will remain an important part of the energy mix for decades to come."[90] The Minister argued that since the IEA predicted oil and gas would fill more than half the world's energy needs at least until 2035, New Zealand should take the opportunity to exploit the country's undiscovered resources. Ironically Bridges's speech coincided with the release of the IPCC's fifth climate change report, warning of the need to reduce reliance on fossil fuels. At the time, Bridges was also associate minister for Climate Change. The juxtaposition of the two events gave the impression the Minister was suggesting New Zealand join the rush to cash in on the deep-sea and unconventional petroleum boom while abrogating its responsibility to future generations to address climate change.

Environmental groups and opposition parties weren't impressed. Forest and Bird's Kevin Hackwell warned that the government's "garage sale" of deep-sea and onshore drilling permits could come at a huge cost to New Zealand's natural heritage through oil spills and contaminated sites.[91] Citing the IPCC's report, Hackwell said "the last thing we need is more fossil fuel development." The New Zealand Greens party concurred as did Greenpeace, arguing that "by opening up whole new areas to drill and mine for oil . . . the government had shown that it was ignoring the science on climate change."[92]

At the October 2014 Petroleum Summit, Minister Bridges repeated his stock rhetoric about the government being committed to "developing the country's resources in a safe and environmentally responsible way." He enumerated his government's accomplishments over the past three years, including establishing a "world class health and safety regime," and ended by foreshadowing further reforms to the RMA.[93] Midway through his presentation, protesters broke into the auditorium, shouting and waving banners calling for an end to fossil fuels. After they were evicted and order was restored, Bridges attempted to answer the protesters with some stock lines about the country's dependency on oil and gas: "We can't just turn off the tap. . . . As we transition to a lower carbon economy we will continue to need petroleum resources in the near future."[94] The government had no comprehensive transition plan at the time.

In December 2014 (following National's re-election) Bridges announced the successful permit bidders for the annual block offer that year and claimed that "public concerns in this area have decreased in the last little while. I think that's because the government has substantially lifted the rules in this area." Protesters against oil and gas exploration clearly disagreed. In March 2016 the minister announced a new set of blocks for auction, emphasizing (in response to criticism of RMA reforms) that it wasn't about "development at any cost." The government, he said, had built a "world-class regulatory framework" for oil and gas development while ensuring "our unique environment is protected."[95] Whether the environment was in fact being protected and natural resources managed sustainably was a moot point. Bridges knew what he was doing. This was a considered and consistent campaign of propaganda and public indoctrination over a period of several years.

Subverting Community Power: From Citizens to Stakeholders

As noted earlier, central and municipal governments often cooperate with oil and gas companies to facilitate corporate engagement strategies. One useful mechanism is the "collaborative stakeholder partnership" or citizen advisory group. Karliner argues such mechanisms are a ploy by which *citizens* are transformed into *stakeholders* in the corporate enterprise. In practice stakeholder partnerships often have more to do with collusion and co-optation than "collaboration." Agreement to participate constitutes acceptance of a process that subsumes local values and development aspirations within a wider corporate and government agenda, which ultimately is part of the process of globalization.

With the National government's introduction of a block auction process and Ministerial promises that officials would work closely with E&P companies who obtained a permit, NZPaM was keen to identify ways to help companies manage their engagement with Māori and local communities. The importance of stakeholder engagement had been underscored at recent Petroleum Summit conferences, and permit holders were required to report to NZPaM on their engagement activities.

One of the first places the idea of a "stakeholder forum" was floated was in Gisborne on the East Coast. In September 2012, NZPaM's Iwi Liaison officer along with a resource management consultant and an MBIE official met confidentially with Gisborne District councilors and iwi representatives and raised the issue of establishing a local stakeholders group. According to a district councilor I interviewed, the Wellington officials suggested such a group could discuss local concerns, commission reports and work toward building a consensus on oil and gas development in the region. It seemed fairly obvious to this councilor that the stakeholder partnership was ultimately intended to facilitate petroleum industry development.[96]

Six months later NZPaM officials made another attempt, this time approaching the Horizons Regional Council whose region stretched from Palmerston North to the East Coast. TAG Oil was already engaged in exploratory drilling near Dannevirke, and NZPaM had advised the council of two additional permit applications to prospect offshore in the region. In February 2014 NZPaM met with the council's Strategy and Policy Committee to provide a briefing on these applications. They took the opportunity to pitch the idea of a "regional stakeholder forum . . . including local authorities, conservationists, landowners, farmers, and potentially industry."[97] NZPaM officials suggested the forum might carry out research and aim to produce a report in twelve months, outlining any problems and solutions related to oil and gas operations in the region. Under questioning by councilors, they were unable to guarantee industry representatives would commit to being involved. The proposal was not followed up by the Council, and the forum never eventuated.

NZPaM has also played an important role in developing procedures and tools to help the petroleum industry engage "appropriately" with Māori, and secure iwi agreements to cooperate with the industry rather than obstruct land access and petroleum development. According to the agency's Chief Advisor Māori, Mahanga Maru, NZPaM worked with Ngāti Ruanui (a Taranaki iwi with an MOU with an oil company) to draw up a set of iwi consultation guidelines for permit holders. They also developed several Crown Minerals Act protocols as well as energy and resource accords with iwi to promote "a new type of relationship to achieve mutually beneficial outcomes."[98]

Agenda-Driven Funding: Buying International Investment and Community Acceptance

Governments in "frontier" petroleum countries have introduced a variety of financial incentives to attract investment from international oil corporations. They have also put in place different carrot-and-stick funding measures to defuse environmental activism and gain the acquiescence if not support of local communities and indigenous people to oil and gas development.

Petroleum production and consumption subsidies total somewhere between $600 and $700 billion globally, roughly 1 percent of world GDP.[99] Critics chastise governments for providing tax breaks to one of the most profitable industries in the world at a time it is becoming increasingly urgent that the world cut fossil fuel emissions to combat climate change. New Zealand government leaders are having to face the growing dilemma of on the one hand playing a lead role internationally in lobbying against petroleum subsidies,[100] while on the other competing with other frontier countries for investment in petroleum exploration and production by offering subsidies and tax breaks.

The Clark Labour government introduced some $6 million in tax breaks and production subsidies to the oil and gas industry. Minister of Energy and Resources Simon Bridges claimed National was "not a government of subsidies or using our cheque book to make things happen."[101] Yet during the first three terms of a National-led government, petroleum industry support increased eightfold. By 2014 the New Zealand taxpayer was subsidizing the industry to the tune of $46 million in tax exemptions. The World Wildlife Fund (WWF) carried out an investigation and estimated that New Zealand fossil fuel subsidies of all types totaled close to $85 million.[102] Greenpeace accused Minister for Climate Change Tim Groser of going around the world championing subsidy reduction while New Zealand's carbon emissions kept going up along with its petroleum industry subsidies. WWF said the government was clearly pursuing a contradictory policy. Groser dismissed the criticisms, calling WWF's report "flawed."

Another tool to promote and support the petroleum industry has been agenda-driven funding. PEPANZ has been influential in establishing the case for targeted funding. In a 2013 briefing paper to government, they argued that international exploration companies would be discouraged from coming to New Zealand if they had to pay for geological survey data that was (supposedly) free in other countries. The government got the message. Just before the September 2014 election, Steven Joyce in his capacity at Minister of Science and Innovation announced that government agency GNS Science had won a $2 million contract to produce a series of freely available (taxpayer funded) digital maps and a comprehensive database "to help exploration companies pinpoint prospective areas to explore for oil and gas in New Zealand's territory." The contract was in line with the government's clamp-down on "non-essential research" and push to prioritize business and development-related research.

In September 2015 GNS Science, the Institute of Environmental Science and Research (ESR), and four New Zealand universities were awarded over $12 million from government's newly created Energy and Minerals Research Fund for geological research aimed at "providing a basis for future oil and gas discoveries" and "attract(ing) more exploration investment here."[103] The main users would be the government and the petroleum industry, who would receive the data for free. The project would also "have *a strong community engagement and industry promotional aspect, which would be funded by the grant.*"[104] The researchers planned to meet with iwi and other stakeholders to explain the "technical aspects" of petroleum exploration and "foster informed dialogue about exploration activity [apparently avoiding 'production' and its impacts] . . . leading to improved understanding of the risks, consequences and scale of potential benefits." Since oil and gas development was not open for public debate, this was clearly a PR exercise on behalf of the government and the industry. Perhaps not surprisingly, the research project included $400,000 in what was euphemistically termed "co-funding" from four international oil companies: Anadarko, ExxonMobil, OMV, and Shell. These companies were already operating in New Zealand so they obviously had no need of enticement. They would however benefit from the research findings and the PR activities of the researchers.

The government also undertook a range of taxpayer-funded promotional activities and spending. Ministers and officials regularly hosted international oil industry executives for dinners, site visits and other events to encourage overseas investment. Government agencies co-sponsored conferences and seminars of significance to their sphere of policy-making.

After National came to power, promotional spending increased from $100,000 to $200,000 a year, which Minister of Energy and Resources Simon Bridges said was a "good investment" for a $2 billion return in taxes and royalties. Critics accused of him of going overboard with taxpayer money in 2011 when promotional spending almost doubled.

The 2011 Petroleum Summit cost taxpayers $237,000 to host, substantially more than two previous conferences.[105] The blow-out was partially due to scheduling the Summit so it coincided with the Rugby World Cup which was held that year in New Zealand and Australia. In an attempt to leverage off the event and attract large overseas oil companies, NZPaM GM Kevin Rollins, previously an Oklahoma oil man himself, arranged for ten oil company executives to be flown to New Zealand and hosted at taxpayer expense.[106] Costs included meals at two top restaurants, a corporate box at the World Cup, a yacht trip, and a winery tour. [107]

In his book *Dirty Politics*, investigative journalist Nicky Hager claimed that in order to silence civil society opponents, the Key government "cut public funding for a wide array of organisations that represented and advocated for communities."[108] There is evidence supporting this claim and indicating that the policy extends to environmental and recreational organizations who criticize government policy and oppose RMA reforms. Under previous governments, community organizations were prohibited from using funding for commercial purposes or political campaigning. During the National government's second term, further reviews were undertaken and the policy tightened. Henceforth, service contracts or public grants could not be used for "advocacy," which was defined broadly to include all forms of lobbying, public statements and political activism. Groups who engaged in such advocacy would lose their funding.[109] The effect if not the intent was to gag community organizations. A Victoria University study released in 2014 showed over half of the groups interviewed feared losing their funding if they criticized government policies or raised issues important to their communities.[110] Some organizations reported receiving threats from officials or ministers, and having gag clauses inserted into their contracts. The researchers pointed out that government relied on feedback from community groups about local needs, and their study showed how "the changing relationship between the state and the community and voluntary sector has resulted in the democratic voice of the sector being heavily constrained."[111]

Environmental and recreational organizations were also targeted. In May 2013 the Minister for the Environment cut funding to the Environmental Defence Society, an environmental law organization that provided free advice and assistance on the Resource Management Act to individuals, groups, and iwi.[112] They had occasionally been critical of government policies. The minister justified the cuts, saying her ministry had reviewed how advice was provided to external stakeholders "in light of changes to the RMA." EDS was forced to seek charitable donations in order to continue their activities. The following year, the same Minister cut funding for environmental organizations and community groups who sought to lodge an RMA appeal. The

Environmental Legal Assistance fund was no longer available for High Court appeals and would not be available for hearings or appeals under the new EEZ Act.

Meanwhile the media had a field day with revelations that Minister for Conservation Nick Smith had threatened to take away Fish and Game New Zealand's statutory powers if they didn't stop criticizing the government.[113] Fish and Game was established under the Conservation Act as an independent body tasked with administering hunting and fishing licenses and advocating for the quality of the country's lakes and rivers. The organization's public campaigning against RMA reforms, expanded irrigation, and agricultural pollution of waterways brought it into direct conflict with government policy. At a meeting with environmental and outdoor recreation groups in July 2014, Minister Smith reportedly warned he would "tweak" Fish and Game's legislation if they didn't rein in their campaigning, which he claimed was threatening economic growth.[114] Dr. Smith subsequently denied he had threatened the organization because of its advocacy activities.

Local Government's Gatekeeping and Advocacy Roles

District and regional councils have a responsibility in legislation for involving their citizens in planning, review of expenditures and services, and natural resource management decisions.[115] Where an activity is not permitted as of right under a district or regional plan, a resource consent must be obtained. Each application must include an assessment of effects of the proposal on the environment. That means petroleum companies must submit environmental impact and mitigation plans to local authorities which are available for public scrutiny. This may be of little relevance unless councils decide to invite public comment on a consent application.

Local authorities have come under increasing pressure from the National government to cooperate with their efforts to promote regional growth and boost primary-sector exports, including oil and gas development. Technical provisions are available under the RMA that can be used if councils choose to "gate keep" on behalf of oil and gas development. Some councils have interpreted these provisions as restricting submitters on a consent application to those who according to the act are "directly affected" by the environmental effects. Others who may have a general interest, concern and/or evidence to present opposing an oil or gas project are excluded. A regional council may simply block public comment on an application by determining on the basis of the applicant's environmental impact statement that the impacts are "less than minor" under the RMA and grant consent without a hearing. The Parliamentary Commissioner for the Environment criticized this practice. In her

2014 report on regulating oil and gas development[116] she called on councils to re-classify all petroleum exploration and production as a "discretionary" activity so the public could be consulted, cumulative impacts could be considered, and conditions attached if the consent was eventually granted.

Iwi consultation can also be used as a gatekeeping mechanism if a majority of councilors are predisposed to oil and gas exploration. Most local authorities have consultation protocols and working agreements with iwi and hapū (tribes and subtribes). In practice councilors and staff only deal with a handful of tribally appointed leaders. Resource consent applications affecting Māori land are seldom discussed at tribal hui, though oil representatives may arrange to brief members about their plans. Individuals or factions within a hapū who object to oil and gas exploration in their rohe (district) can make their views known to their leaders, local councilors, and the media but the municipal authority is only required to take account of the views of recognized tribal representatives.

At least in principle, local councilors and mayors engage in a constant juggling act between promoting economic development, regulating and controlling land-use, and considering all the ramifications of new developments for the well-being of their district. In practice it is not unusual for individual councilors and sometimes whole councils to become cheerleaders for an industry like oil and gas which promises huge benefits to the community. For instance, cheerleading became the subject of council debate and community criticism in March 2015 when Shell Todd Oil Services (STOS) applied to the EPA for a marine consent to continue production at the company's offshore Maui gas field. The Taranaki Regional Council (TRC) policy and planning committee was asked to approve a submission, drafted by council staff, supporting STOS's application. The Council's prime responsibility is managing the use, development, and protection of the region's natural and physical resources. Councilor Maurice Williamson criticized the draft submission, arguing it was not neutral as required of the regional council as an environmental regulator. Instead, it supported STOS's application and highlighted the economic benefits of the company's gas operations. Prominent local activist Sarah Roberts echoed Williamson's comments in a submission to the committee, accusing the council of "cheerleading."[117] TRC Director of Resource Management Fred McLay (an ex-employee of STOS) defended the submission. After a majority of councilors approved the submission, McLay presented it in person at EPA hearings several weeks later. There have been similar cases of municipal council cheerleading for petroleum development in other parts of the country.[118] Local communities understand how the game is played, and most elected representatives try to at least give the appearance of listening to and taking account of the views of all their constituents, even if they are biased toward development "at all costs."

There have only been a few instances to date where the propriety of the relationship between the extractive sector and local politicians has come in for public scrutiny. In 2012 the mayor of the Far North District Council Wayne Brown received council support for a $16,000 trip to a mining trade show in Toronto followed by a petroleum and minerals exploration symposium in Vancouver. The trip was in cooperation with government agencies NZPaM, Trade and Enterprise (NZTE), and GNS Science who were sending their own representatives. Local critics and the New Zealand Greens party accused the mayor of promoting mining in Northland without proper public consultation and misusing ratepayers' money. [119] At the time, Brown was chairman of the Explore Northland Minerals Group, a "collaborative partnership" organized and funded by the Ministry for Economic Development which included several mining companies. The controversy damaged Brown's standing with the electorate, and he was voted out of office at the next local body election. Following his loss, Brown used his business contacts to become a community/ iwi engagement consultant for the mining industry.

Royalties from oil and gas exploration and production in New Zealand go into the central government's general fund and are used for benefit of the whole country. With government plans to greatly expand the industry, local politicians began lobbying for a share of royalties to be allocated to district councils to pay for the impacts of oil and gas development on local infrastructure and services. In the run up to the 2014 election Local Government New Zealand (LGNZ), representing local authorities throughout the country, saw an opportunity to advocate for royalty sharing. When New Zealand First MP Winston Peters raised the issue in March, LGNZ jumped on the bandwagon agreeing that a portion of royalties from all mining activity should be returned to local and regional councils to cover community planning, infrastructure, and environmental management costs. [120] Ironically, by extolling the benefits of the industry to communities while calling for royalty-sharing, LGNZ was inadvertently acknowledging that oil and gas development *did create* economic, social and environmental costs for communities that had to be managed.

The election debate over royalty sharing again surfaced among Northland MPs and local government representatives a few weeks later. LGNZ's chief executive Malcolm Alexander insisted in a Radio NZ interview that oil and gas royalties could help regions "build a future" when the boom came to an end. [122] Embarrassingly for Alexander, a month later Tararua District Council's chief executive Blair King was reported in the media as warning his councilors that dependence on royalties could actually exacerbate a boom and bust cycle. According to a staff research paper, "money from royalties can really benefit a community, but when the tap turns off it leaves a community far less resilient."[122]

District and regional councils have been quick to reassure residents that they have robust processes and regulatory systems in place to ensure drilling and production are done properly and safely. The reality, as the Parliamentary Commissioner for the Environment disclosed in her two fracking reports, is that there are significant problems yet to be addressed if the industry is to be allowed to expand.[123] Although the Taranaki Regional Council has been held up by the government and the petroleum industry as setting the regulatory benchmark in New Zealand, the TRC's track record tells a different story. According to the PCE in her 2014 report, the regulatory regime of the TRC and several Taranaki district councils was "clearly out of step with best practice." Among other problems, the commissioner pointed to the lack of need for a consent to drill oil and gas wells, a lack of a comprehensive checklist of environmental considerations for assessing applications, failure to monitor effects against consent requirements, no assessment process for dealing with the cumulative effects of an expanding industry, and allowance of unproven methods of waste disposal.[124] This was a scathing indictment of the country's supposedly leading regional resource regulator.

An investigation by *Live Magazine*[125] confirmed the commissioner's findings and uncovered even more problems across the Taranaki region as a whole:

- Contrary to provisions in the RMA, petroleum consent applications to district councils weren't being considered in conjunction with discharge consents by the TRC. It meant that the TRC was treating all applications the same and allowing discharge of contaminants without knowing precisely what hazardous substances were being used.
- Petroleum companies sometimes gave contradictory information to different councils for the same activity; councils did no cross-checking but accepted the data at face value.[126]
- The public was not being given the opportunity to participate in consent hearings because councils routinely found the effects in company environmental impacts statement (EIS) documents were "less than minor." The criteria for such judgments were rarely spelled out nor were EIS documents independently reviewed to see if they were factual and met RMA requirements.[127]
- Each district council had a different set of rules and mitigation measures for oil and gas companies to comply with.

- Councils did not consistently monitor drilling and production operations for compliance with regulations or environmental damage; staff were often tardy in following up complaints.[128]

The *Live Magazine* investigation also identified overly cozy and potentially compromising relationships between industry representatives and consultants, government officials and Taranaki regulatory staff.[129]

CONCLUSION

Under successive National coalition governments, there has been a shift in economic policy toward commodity-led export growth and expansion of extractive industries. The government has been able to rely on powerful private-sector interests including the oil and gas industry for support and active involvement in implementing this agenda. To advance their Business Growth Agenda and remove "impediments" to petroleum industry expansion, the National government embarked on a program of legislative reform that resulted in centralizing economic development decision-making, weakening participatory local government, and dismantling of the sustainable development framework a succession of governments had put in place.

These legislative moves were developed over time and presented to the public as essential for promoting business growth and prosperity in an "environmentally responsible way." The government sought the cooperation and direct involvement of the petroleum industry and its consultants in developing policy (e.g., the 2009 Petroleum Strategy), shaping legislation, and promoting its "environmentally responsible" policies and "safe, world class" industry operations to the public. To support these policy and legislative moves, National developed what in effect constituted a promotional "program" in which the Ministers of Energy and Resources, Economic Development, and Environment played key roles. The informal playbook involved carefully contrived rhetorical spin, "official" information dissemination and carefully orchestrated policy consultation exercises (*with the exception of energy policy*). Critics claimed that the government/petroleum industry relationship occasionally bordered on collusion, as in regard to ministerial meetings with industry representatives immediately prior to urgent EEZ legislation banning protests near offshore petroleum operations.

As a result, a shift in institutional protocols and relationships began to occur toward closer collaboration between government ministers, bureaucrats, petroleum executives, and consultants to clear impediments to industry expansion, promote petroleum development to the New Zealand public, and shut down environmentalist critics and local activists. The new institutional arrangements became especially apparent during the prolonged struggle to make "sweeping changes" to the RMA.

At the same time, increasing public (and business) concerns over preserving New Zealand's clean green image, protecting the natural environment, and addressing climate change—heightened by the government's determined efforts to promote oil and gas development—forced National to rethink its confrontational stance at least toward mainstream environmentalists in order to maintain the support of Middle New Zealand. The rapidly changing context for political strategizing also exacerbated tensions between the conservative wing of the National Party who pressed for more radical legislative reform to stimulate business growth, and the Blue-Green faction who saw the need to balance growth with environmental protections and "responsible resource management." Although government ministers initially rebuffed proposals from the Parliamentary Commissioner for the Environment and Pure Advantage to consider "alternative governance arrangements" for addressing climate change and transitioning to a sustainable, low-carbon economy, by the end of 2015 with the COP21 climate summit immanent, the government toyed with the idea of establishing a national cross-sector forum and then opted for a ministerial working group to ensure government control of the planning process.

Taking a cue from the success of the collaborative Land and Water Forum and nudging from the Blue-Green faction of MPs, the government decided to adopt a more corporatist approach to engaging with (and hopefully co-opting) moderate environmental organizations and outdoor recreational groups, even if this meant occasionally chastising a partner group for getting out of line. Prime Minister Key met with environmental groups following the Northland bi-election, promising a "less confrontational approach" to achieving RMA reforms and "sustainable growth." When Finance Minister Bill English announced the blue-skies review by the Productivity Commission in late 2015, he stressed that the Commission would seek to involve all sectors of society in a deliberative process.

But in general the government adopted a more hardline approach to the community sector, clamping down on "advocacy" and passing legislation that reduced public participation in local governance and resource management while maintaining rhetoric in the social services sector about communities and government working together. This was certainly in line with international research regarding how states influenced by extractive industry often tilt toward authoritarian strategies in relation to civil society. Extreme environmental organizations like Greenpeace, Climate Justice Taranaki, and local anti-fracking groups—all of whom were highly critical of government's economic and environmental policies—were excluded (or excluded themselves) from collaborative government consultations, working groups and local "stakeholder partnerships." Their criticisms and protests were generally dismissed out of hand by government and industry spokespersons. Institutionally they were left outside the pale and marginalized from influencing policy development and legislative reforms. They did however continue to inform and influence public debate over government's policies with demonstrations, submissions and media releases challenging the government's "factual" information, pro-industry propaganda, and climate change record. As such they continued to pose a threat to National's efforts to capture Middle New Zealand and avoid a new groundswell of popular environmentalism.

In the next chapter we will examine the strategies the oil and gas industry in New Zealand created or adopted from overseas in order to promote itself to the political establishment and the general public, while defending itself against the criticisms and actions of environmental organizations and local anti-petroleum activists. Most of these strategies assumed and relied upon the close collaboration of the New Zealand state apparatus and key ministers.

NOTES

1. See the open letter from 144 health practitioners, scientists, and professional health organizations to Hon Andrew Cuomo, governor of the State of New York, and Howard Zucker, acting commissioner of the New York State Department of Health, May 29, 2014. http://concernedhealthny.org/letters-to-governor-cuomo/. The submission was influential in the governor's decision to ban fracking in New York State.

2. Michael King, *The Penguin History of New Zealand* (Auckland: Penguin, 2003), 442.

3. King, *Penguin History of New Zealand*, 443.

4. In July 2001, Cabinet agreed that the principles of sustainable development should underpin all of the government's economic, social, and environmental policies. See New Zealand Parliamentary Commissioner for the Environment, "Creating Our Future: Sustainable Development for New Zealand" (New Zealand Government, Wellington, June 2002), https://www.smartgrowthbop.org.nz/media/1345/Creating-Our-Future-June-2002.pdf.

5. New Zealand Petroleum and Minerals, 2014, www.NZPaM.govt.nz, emphasis added.

6. McDouall Stuart, "Stepping Up: Options for Developing the Potential of New Zealand's Oil, Gas and Minerals Sector," Report to the Ministry for Economic Development (New Zealand Government, Wellington, June 2009), http://www.mbie.gov

t.nz/info-services/sectors-industries/natural-resources/oil-and-gas/petroleum-expert
-reports/McDouall-Stuart.pdf.

7. Gerry Brownlee, "Unlocking New Zealand's Petroleum Potential," Media
release, New Zealand Government, Wellington, November 18, 2009, https://www
.beehive.govt.nz/release/unlocking-new-zealand039s-petroleum-potential.

8. David Coull, "Significant Period of Change and Regulatory Reform in the
New Zealand E&P Sector," *Who's Who Legal* (June 2014), www.whoswholegal.com.

9. Ministry of Business, Innovation and Employment, "Developing Our Energy
Potential: The New Zealand Energy Strategy 2011-2021 and the New Zealand Energy
Efficiency and Conservation Strategy" (New Zealand Government, Wellington, August 2011), http://www.mbie.govt.nz/info-services/sectors-industries/energy/energy
-strategies/documents-image-library/nz-energy-strategy-lr.pdf.

10. MBIE, "The New Zealand Energy Strategy 2011-2021."

11. MBIE, "The New Zealand Energy Strategy 2011-2021," Introduction.

12. See Peter Benson and Stuart Kirsch, "Corporate Oxymorons," *Dialectical
Anthropology* 34 (2010): 45–48, doi 10.1007/s10624-009-9112-y.

13. Rod Oram, "Growth at Any Price Foolish," Opinion, *Sunday Star-Times*, November 18, 2012, accessed May 17, 2015.

14. "Government 'Mockery' of Green NZ," *Stuff News*, July 9, 2013, accessed
November 3, 2014.

15. Information provided by the Royal Society of New Zealand, see http://www
.royalsociety.org.nz/events/ten-by-ten/ten-by-ten-climate-change, accessed July 7,
2016.

16. Parliamentary Commissioner for the Environment, "New Zealand's Contribution to the New International Climate Change Agreement: Submission to the Minister
for Climate Change Issues and the Minister for the Environment," New Zealand Government, Wellington, June 3, 2015, 6, http://www.pce.parliament.nz/. For a similar
argument see Sustainability Aotearoa New Zealand, *Strong Sustainability for New
Zealand: Principles and Scenarios*, SANZ, Auckland, 2009, 31, http://www.earths-
limits.org/resources/2014/5/24/strong-sustainability-for-new-zealand.

17. Greenpeace New Zealand, *The Future is Here: New Jobs, New Prosperity,
and a New Clean Economy*, Auckland, February 2013, http://www.greenpeace.org/
new-zealand/en/campaigns/climate-change/The-Future-is-Here/. World Wildlife
Fund-New Zealand, *Fossil Fuel Finance in New Zealand, Part 1: Government
Support*, Auckland, 2013, http://awsassets.wwfnz.panda.org/downloads/wwf_fossil
_fuel_finance_nz_subsidies_report.pdf. The Royal Society of New Zealand, *Facing the Future: Towards a Green Economy in New Zealand*, Wellington, 2014,
http://royalsociety.org.nz/expert-advice/papers/yr2014/greeneconomy/. Sustainable
Business Council, *Vision 2050: The Report—Business as Usual is not an Option*,
Auckland, 2012, http://www.sbc.org.nz/__data/assets/pdf_file/0015/57120/Vision
-2050-NZ-Report-2012.pdf.

18. MBIE, "Building Natural Resources" (New Zealand Government, Wellington, December 2012), 20, http://www.mbie.govt.nz/info-services/business/business
-growth-agenda.

19. Ministry for the Environment, "History of RMA Reforms" (New Zealand Government, Wellington, 2013), www.mfe.govt.nz.

20. Ministry for the Environment, "Technical Advisory Group 2 Report" (New Zealand Government, Wellington, February 2012), 10–12, https://www.mfe.govt.nz/sites/default/files/tag-rma-section6-7.pdf.

21. The assumption underlying the revised principles was that "resources" were an asset to be used, exploited and 'managed' for economic and social benefit. Such assumptions sat uncomfortably with principles in the RMA recognizing Māori values and preserving natural landscapes.

22. TAG2 appears to have been wrong in law. On April 17, 2014, in the case of the *Environmental Defence Society v NZ King Salmon Ltd*, the Supreme Court declined proposals for several fish farms in the Marlborough Sounds in part because the project failed to pay sufficient heed to the requirement in the RMA to *protect* outstanding landscapes.

23. Fish and Game New Zealand, "Fish and Game Opposes 'Attack' on RMA," Media release, Wellington, September 11, 2012, http://www.fishandgame.org.nz/newsitem/fish-game-opposes-attack-rma.

24. Eugenie Sage, "Report Major Assault on the RMA," New Zealand Greens, Media release, July 5, 2012, https://home.greens.org.nz/press-releases/report-major-assault-rma.

25. Amy Adams, "Improving New Zealand's Resource Management System," Media release, March 1, 2013, New Zealand Government, Wellington, https://www.beehive.govt.nz/minister/amy-adams.

26. Environmental Defence Society, "EDS Position Paper on RMA Reforms," Newsletter, September 3, 2013, accessed September 13, 2013, www.eds.org.nz.

27. DLA Phillips Fox, "Implications of Proposed Changes to Part 2 of the RMA," Report to the Environmental Defence Society, September 2, 2013, para 9, www.eds.org.nz.

28. Environmental Defence Society, "EDS Critical of Proposed RMA Reforms," Newsletter August 10, 2013, accessed September 13, 2013, www.eds.org.nz.

29. Sir Geoffrey Palmer QC, "Protecting New Zealand's Environment: An Analysis of the Government's Proposed Freshwater Management and Resource Management Act 1991 Reforms," Report commissioned by Fish and Game NZ (Fish and Game NZ, Wellington, September 11, 2013), 3, emphasis added. Sir Geoffrey frequently refers to the National Government's "growth-at-any-cost" approach.

30. Palmer QC, "Protecting New Zealand's Environment," 23, emphasis added.

31. Ministry for the Environment, "Summary of Submissions on 'Improving Our Resource Management System,'" New Zealand Government, October 2013, www.mfe.govt..nz.

32. Patrick Smellie, "What Will be Done with The RMA?" *National Business Review*, October 23, 2014, accessed January 29, 2015.

33. "Increased Support for Nats 'Sends Clear Message,'" *Radio New Zealand*, 22 September 2014, accessed September 22, 2014.

34. Pattrick Smellie, "What Will be Done with the RMA?"

35. Emphasis added. Evidently the "environmentally aware" minister shared his colleagues' expedient view that certain developments should proceed even though they caused unavoidable damage ("necessary harm"?) to the environment.

36. Environment and Conservation Organisations of Aotearoa New Zealand. ECO was founded in 1972, during the Save Manapouri Campaign and currently has more than fifty member organizations, see www.eco.org.nz.

37. "Northland Loss Would Make RMA Reform Difficult," *TV3*, March 24, 2015, accessed March 26, 2015.

38. "Time to 'Rip Up' RMA Reforms in Wake of Defeat," *New Zealand Herald*, March 30, 2015, accessed June 14, 2015. "Government to 'Rip Up' RMA Plans," *Radio New Zealand*, March 30, 2015, accessed June 14, 2015, emphasis added.

39. "'Post-RMA' Process Looms as National Seeks New Urban Planning Report," *National Business Review*, November 2, 2015, accessed November 3, 2015, emphasis added. See also *Interest.co.nz*, "English Launches 'Blue Skies' Productivity Commission Review into Resource Management, Local Government and Land Transport Management Acts to Transform Urban Planning and Tackle Housing Affordability," November 2, 2015, accessed November 3, 2015.

40. "English launches . . . ," *Interest.co.nz*, emphasis added.

41. Political commentator Pattrick Smellie reached a similar conclusion, predicting a five-year program of RMA/planning reform. See "Pattrick Smellie: First Rule of RMA Reform: It Never Ends," *Stuff News*, December 9, 2015, accessed December 10, 2015.

42. In the United States, landowners own mineral rights under their land, unless they have not been included in the original deed title.

43. Gerry Brownlee and Kate Wilkinson, "Time to Discuss Maximising our Mineral Potential," Media release from the Minister of Energy and Resources, and Minister of Conservation, New Zealand Government, Wellington, March 22, 2010, https://www.beehive.govt.nz/release/time-discuss-maximising-our-mineral-potential.

44. Gundars Rudzitis and Kenton Bird, "The Myth and Reality of Sustainable New Zealand: Mining in a Pristine Land," *Environment Magazine*, September/October 2011, http://www.environmentmagazine.org/Archives/Back%20Issues/2011/November -December%202011/Myths-full.html.

45. Thomas O'Brien, "Fires and Flotillas: Opposition to Offshore Oil Exploration in New Zealand," *Social Movement Studies: Journal of Social, Cultural and Political Protest* 12, 2 (2013): 221–226.

46. Nathan Argent, "Our Right to Protest Is Vital," Greenpeace New Zealand, Annual Impact Report, Auckland, 2013, http://www.greenpeace.org/new-zealand/ Global/new-zealand/P3/publications/other/Greenpeace%20NZ%20Annual%20 Impact%20Report%202013.pdf.

47. Interview with Stuart Nash, Labour spokesperson on Energy, December 2, 2015. The interviewee was advised the discussion was being recorded and consented to the information provided being available for possible publication. Labour leader David Cunliffe told TV3 on February 6, 2014, that a Labour government would review existing oil and gas permits and regulations, further tighten safety and environ-

mental regulations, and make "most" future consent applications publically notifiable rather than non-notifiable.

48. Greenpeace New Zealand, "Energy Minister Challenged to Release Full Meeting Details over Misleading Parliament Accusation," Media release, June 4, 2013, www.greenpeace.org.nz.

49. Greenpeace New Zealand, "Letter to the Honourable Mr Simon Bridges, regarding Amendments to the Crown Minerals Bill," media release, June 4, 2013, www.greenpeace.org.nz.

50. David Coull, "Significant Period of Change."

51. Bell Gully, "Draft Regulations for Exploratory Oil and Gas Drilling Under the EEZ Act," *Lexology*, December 19, 2013.

52. Gary Taylor, "Secretive Regime on Drilling Dangerous," Opinion: *New Zealand Herald*, January 13, 2014, emphasis added.

53. Argent, "Our Right to Protest."

54. Amy Adams, "New Regulations for Exploratory Drilling," Media release, New Zealand Government, Wellington, February 27, 2014, https://www.beehive.govt.nz/release/new-regulations-exploratory-drilling.

55. Kate Gudsell, "Govt Accused of Breaking Marine Promise," *Radio NZ News*, March 9, 2016, accessed March 10, 2016. "EDS Supports New Marine Protected Areas Act but Bizarre Omission Needs Fixing," Media release, Environmental Defence Society, March 9, 2016, http://www.eds.org.nz/our-work/policy/media-statements/media-statements-2016-1/media-release-eds-supports-new-marine-protected/.

56. "Poll: Marine Reserve Reform Should Cover EEZ," *New Zealand Herald*, July 31, 2016, accessed July 31, 2016.

57. The author was a senior advisor with the Department of Internal Affairs from 2001 to 2006, and a member of the inter-agency officials group responsible for rolling out the LGA 2002.

58. Local Government Act 2002, Part 1 Section 3 (d) of the Purpose Statement (Wellington, New Zealand Government, 2002), 9.

59. Department of Internal Affairs, *Smarter Government, Stronger Communities: Towards Better Local Governance and Public Services* (New Zealand Government, Wellington, February 2, 2011), https://www.dia.govt.nz/Resource-material-Our-Policy-Advice-Areas-Smarter-Government-Stronger-Communities.

60. Rodney Hide, "Local Government Act 2002 Amendment Act: Decisions for Better Transparency, Accountability and Financial Management of Local Government" (Department of Internal Affairs, Wellington, November 26, 2010), http://www.baybuzz.co.nz/wp-content/uploads/2010/08/HideLGA.pdf.

61. Minister for Local Government, "Better Local Government: Opportunities to Improve Efficiency," Paper to the Cabinet Economic Growth and Infrastructure Committee (Department of Internal Affairs, July 2013), https://www.dia.govt.nz/Resource-material-Our-Policy-Advice-Areas-Smarter-Government-Stronger-Communities.

62. Department of Internal Affairs, "Report of the Local Government Efficiency Taskforce" (New Zealand Government, Wellington, December 2012), https://www.dia.govt.nz/pubforms.nsf/URL/Local-Government-Efficiency-Taskforce-Final-Report-11-December-2012.pdf.

63. DIA, Efficiency Taskforce, paragraph 109, emphasis added.

64. Oliver Hartwich, "A Global Perspective on Localism" (Published jointly by Local Government New Zealand and The New Zealand Initiative, Wellington, 2013), 16, http://www.lgnz.co.nz/assets/Publications/A-global-perspective-on-localism.pdf.

65. Hartwich, "A Global Perspective on Localism," 11.

66. *Yes, Prime Minister*, comedy series written by Antony Jay and Jonathan Lynn, London: BBC Television, 1986–1988.

67. "Supersized Staff Level for 'Super Ministry,'" *Stuff News*, May 29, 2014.

68. MBIE, "Economic Contribution and Potential of New Zealand's Oil and Gas Industry" (Occasional paper 12/07, New Zealand Government, Wellington, July 2012), http://www.mbie.govt.nz/info-services/sectors-industries/natural-resources/ oil-and-gas/petroleum-expert-reports/economic-contribution-and-potential-of-new -zealand2019s-oil-and-gas-industry/.

69. MBIE, "Economic Contribution and Potential," 4.

70. MBIE, "Economic Contribution and Potential," 5. Research cited earlier in this discussion casts doubt on such claims.

71. MBIE, "Economic Contribution and Potential," 8.

72. MBIE, "Economic Contribution and Potential," 16. The authors cite Venture Taranaki, an economic development agency that promotes oil and gas development, as the source of this data.

73. MBIE, "East Coast Oil and Gas Development Study" (New Zealand Government, Wellington, March, 2013), 5, http://www.mbie.govt.nz/info-services/sector s-industries/natural-resources/oil-and-gas/petroleum-expert-reports/east-coast-oil -and-gas-development-study/East-Coast-oil-gas-development-study-report.pdf.

74. MBIE, East Coast Oil and Gas Development Study, 5.

75. Steven Joyce and Simon Bridges, "Petroleum and Minerals Sector Boosts NZ Economy," Media release, Wellington, September 2, 2013, https://www.beehive.govt .nz/release/petroleum-amp-minerals-sector-boosts-nz-economy.

76. MBIE, "East Coast Regional Economic Potential Stage 1: Research Review," and "East Coast Regional Economic Potential Stage 2: Economic forecasting and transport and skills implications" (New Zealand Government, Wellington, April 2014), http://www.mbie.govt.nz/info-services/sectors-industries/regions-cities/ regional-growth-programme/east-coast.

77. "Hurry the Plan: Minister," *Gisborne Herald*, March 26, 2016, accessed March 28, 2016.

78. Interview with Josh Adams, NZPaM National Manager Petroleum, December 2, 2015. The interviewee was advised the discussion was being recorded and consented to the information provided being available for possible publication.

79. "Oil Meeting Cancelled," *Marlborough Express*, October 24, 2013, accessed September 18, 2014.

80. "Community Says No to Drilling," *Stuff News*, November 11, 2013, accessed June 25, 2015.

81. Over two hundred people attended the November 8, 2013, hui at Ahipara, Northland, for an NZPaM "consultation." The meeting expressed unanimous opposition to mining and deep-sea drilling, and ejected the officials from the marae.

82. "Northland Regional Council Organising Two Workshops," *Scoop News*, August 25, 2015, accessed August 26, 2015.

83. "Two workshops," *Scoop News*.

84. Edward Herman and Noam Chomsky, *Manufacturing Consent.*

85. See Gary Taylor, "Secretive Regime on Drilling Dangerous," *New Zealand Herald*, January 13, 2014, accessed June 10, 2015: "We are all aware that the Government is keen to encourage oil and gas exploration. Indeed, the Minister of Energy has been embarrassingly effusive in his advocacy for the sector."

86. Simon Bridges, "Speech to Advantage NZ: 2013 Petroleum Conference," Media release, Wellington: New Zealand Government, April 29, 2013, https://www .beehive.govt.nz/speech/speech-advantage-nz-2013-petroleum-conference.

87. Bridges, Advantage New Zealand conference speech, 2013.

88. G Diprose, AC Thomas & S Bond, 'It's Who We Are': Eco-nationalism and Place in Contesting Deep-sea Oil in Aotearoa New Zealand, *Kōtuitui: New Zealand Journal of Social Sciences Online* 11, 2 (2016): 159–173. doi: 10.1080/ 1177083X.2015.1134594

89. Simon Bridges, "Block Offer 2014: Address to the Advantage NZ 2014 Geotechnical Petroleum Forum," New Zealand Government, Wellington, April 2, 2014, https://www.beehive.govt.nz/speech/block-offer-2014-address-advantage-nz -2014-geotechnical-petroleum-forum.

90. "Government Opens Eight Areas for Oil and Gas Exploration," *TV1 News*, April 2, 2014, accessed September 20, 2014.

91. "Huge Oil and Gas "Garage Sale" Comes at Real Cost to Nature," *Scoop News*, April 2, 2014, accessed September 19, 2014.

92. "Continued Hunt for Oil, Gas Angers Greenpeace." *Stuff News*, April 2, 2014, accessed September 19, 2014.

93. Simon Bridges, "Speech to 2014 New Zealand Petroleum Summit," Media release, Wellington: New Zealand Government, October 1, 2014, https://www.beehive .govt.nz/speech/speech-2014-new-zealand-petroleum-summit.

94. "Security Not Tight Enough at Oil Summit," *Taranaki Daily News*, October 1, 2014, accessed October 25, 2014.

95. Daniel Graeber, "New Zealand Offering More Oil, Gas Acreage," *UPI*, March 21, 2016, accessed March 25, 2016.

96. Interview with a local district councilor September 16, 2012. The councilor consented for the information to be used in possible publication on condition of anonymity.

97. "NZPaM Pitches Stakeholder Forum Idea to Horizons Regional Council," *Manawatu Standard*, February 12, 2014, accessed September 29, 2014.

98. Mahanga Maru, "NZPaM's Role and Engagement with Iwi" (presentation to the Symposium on Māori Engagement with Extractive Industry, Faculty of Law, Auckland University, June 12, 2015), http://www.law.auckland.ac.nz/en/about/ events-1/events/events-2015/01/06/symposium-on-maori-engagement-with-extractive -industry--innovati.html.

99. Organization for Economic Cooperation and Development, *Inventory of Estimated Budgetary Support and Tax Expenditures for Fossil Fuels* (OECD, Paris,

2011), http://www.oecd.org/site/tadffss/48805150.pdf. See also Charles Nsobya, *New Vision* Newspaper, April 24, 2015.

100. World Wildlife Fund, *Fossil Fuel Finance in New Zealand.*

101. Richard Harmon, "The Generation X Minister Driving NZ (and His Own Career) Along the Technology Curve," *Politik*, March 3, 2016, http://politik.co.nz/en/content/politics/565/

102. WWF, *Fossil Fuel Finance in New Zealand.*

103. "New Research Programme Aims to Lift Oil & Gas Exploration," *Scoop News*, September 3, 2015, accessed September 5, 2015.

104. MBIE, "Energy and Minerals Research Fund—2015 Science Investment Round Successful Proposals," Media release, New Zealand Government, Wellington, September 3, 2015, https://www.beehive.govt.nz/portfolio/science-and-innovation?page=1.

105. "Oil Executives Treated to Expensive Wine Tasting," *New Zealand Herald*, August 11, 2014, accessed September 18, 2014.

106. "Minister Defends Money Spent on Oil and Gas Promotion," *New Zealand Herald*, July 29, 2014, accessed September 18, 2014.

107. "Oil Companies Hosted During Rugby World Cup," *TV3 News*, July 29, 2014. Also "Spending on Oil Barons Draws Criticism, *New Zealand Herald*, July 30, 2014, accessed September 18, 2014.

108. Nicky Hager, *Dirty Politics: How Attack Politics is Poisoning New Zealand's Political Environment* (Nelson: Craig Potton Publishing, 2014), 74.

109. The author was a member of the New Zealand Lottery Board's Community Sector Research Committee at the time reviews were undertaken by the Board and the Department of Internal Affairs.

110. "Community Groups 'Fear Speaking Out,'" *Radio New Zealand*, November 20, 2014. Also "Community Groups Facing Financial Pressure," Radio New Zealand, November 15, 2014, accessed November 22, 2014.

111. Sandra Grey and Charles Sedgwick, "The Contract State and Constrained Democracy: the Community and Voluntary Sector under Threat," *Policy Quarterly* 9, 3 (August 2013): 3–10.

112. Environmental Defence Society, "Government Cuts Funding to EDS," Media release, Auckland, May 10, 2013, www.eds.org.nz

113. "Minister Nick Smith Threatens Fish and Game," *Radio New Zealand*, July 27, 2014, accessed September 19, 2014.

114. "Smith Threatens," *Radio New Zealand.*

115. E.g. Resource Management Act 1991, Sections 44, 46, and 46A and Schedule 1.

116. Parliamentary Commissioner for the Environment, *Drilling for Oil and Gas in New Zealand: Environmental Oversight and Regulation* (New Zealand Government, Wellington, June 2014), 77.

117. "TRC 'Bias' Slammed in Oil-Gas Hearing," *Taranaki Daily News*, March 23, 2015, accessed March 31, 2015.

118. For examples of cheerleading and attacks on oil drilling critics, see comments by Dannevirke Mayor Rolly Ellis about TAG Oil's operations near his town

in "Tag Oil Natural Gas Found in Dannevirke First Well," *Hawke's Bay Today*, May 17, 2013. See also"Fears Fracking Will Fracture Communities," *Radio New Zealand*, January 31, 2014, accessed June 10, 2014.

119. "Mayor to Lure Mining Firms to Northland," *Stuff News,* February 24, 2012. "Brown Takes Mining Message to Canada," *The Northern Advocate*, February 17, 2012, accessed September 27, 2015.

120. "Sharing Mining Royalties 'Positive for Councils and Communities,'" *Voxy News*, March 21, 2014, accessed September 20, 2014, emphasis added.

121. "Regions Call for Resource Royalties," *Radio New Zealand*, September 5, 2014, accessed September 19, 2014.

122. "Dannevirke: Warning over Oil and Gas Royalties," *Hawke's Bay Today*, October 1, 2014, accessed October 3, 2014.

123. Parliamentary Commissioner for the Environment, "Evaluating the Environmental Impacts of Fracking in New Zealand: An Interim Report" (New Zealand Government, Wellington, November 2012), 76ff.

124. Parliamentary Commissioner for the Environment, "Drilling for oil and gas in New Zealand," 5ff, 21ff, 29ff

125. *Live Magazine*, "Is Taranaki the New Texas?" (Autumn 2012): 8–14.

126. *Live Magazine*, "New Texas?"

127. *Live Magazine*, "New Texas?" 11. See for example consultant Norbert Schaffoener's report to the New Plymouth District Council regarding Greymouth Petroleum's "barely acceptable" risk assessment for its Kowhai B well site.

128. *Live Magazine*, "New Texas?" When a local resident in the vicinity of Todd Energy's Mangahewa C well pad asked for all council environmental assessments after six years of the site's operation, she was told there were none.

129. "A Rock and a Hard Place: Deep Fissures over Fracking," *Nelson Mail*, June 25, 2012, accessed October 8, 2014.

Chapter Four

Petroleum Industry Strategies to Promote and Defend Oil and Gas Development

Economist Geoff Bertram once observed that Kiwis[1] have a propensity to gullibility when dealing with promoters. Promoters have an interest in making the strongest possible case for their business or pet project. "They will be honest and open only in a policy environment where openness and honesty pay and where naked propaganda doesn't pay off. New Zealand is not such an environment."[2]

Around the world the fossil fuel sector is increasingly viewed as a sunset industry. Even leading petroleum executives and PEPANZ's CEO Cameron Madgwick[3] acknowledge the world needs to transition to a clean energy economy, eventually. Other critics argue it is the tobacco industry of the twenty-first century: addictive, harmful, yet prepared to defend itself at all costs from regulatory extermination. In this chapter we will examine some of the promotional and defensive strategies the petroleum industry has employed in collaboration with the New Zealand government to gain acceptance for increased oil and gas development and combat environmental activism. At the end of the chapter we will reflect on the particular historic, political, and social circumstances that have influenced the industry's choice of strategies in this country, and what changes in strategic approach and state/industry and community sector relations might be in store in the near future.

THE OIL AND GAS INDUSTRY PROMOTION NETWORK IN NEW ZEALAND

Earlier we noted Oreskes and Conway's investigation of how the tobacco industry went about manufacturing doubt and attacking scientific research about the harm it did. They found that Big Oil and PR consultants had borrowed some of these same tactics in a concerted effort to promote the petroleum industry, generate a "balanced" energy debate, and undermine the claims of environmentalists. Similarly, when Dunlop and McCright studied the climate denial movement, they found that what appeared to be a spontaneous public debate was in reality an organized campaign by a sophisticated machine comprised of non-profit think tanks, front groups, and "experts" backed covertly by corporate money including the oil and gas industry.[4]

Does such a "machine" exist in New Zealand to promote and defend the oil and gas industry? The answer is yes and no. Most exploration and production companies have someone on their staff responsible for PR or community/ stakeholder relations. But there isn't the extensive system of specialist PR firms, lobbyist consultancies, industry-funded think tanks, and front groups that has evolved in large petroleum-producing countries. Petroleum industry promotion and defense occurs largely through an informal network of personal and organizational relationships between government officials, industry executives, paid lobbyists, individual consultants, and a handful of supportive journalists and bloggers (see figure 4.1). But the network does include some of the biggest corporations and most influential political and business leaders in the country. Based on analysis of business media articles, company and government reports, submissions and briefings to government, company websites, board memberships, LinkedIn relationships, conference programs, and interviews, what seems to bind this network together is a pragmatic value system and broadly shared agenda regarding the continuing importance of oil and gas.

The main components of the New Zealand petroleum promotion network are *the petroleum industry, central and local government, corporate New Zealand, and a handful of conservative trusts and foundations* who receive direct or indirect funding from the petroleum industry. The finance and legal sectors play an important role in advising, supporting, and investing in the petroleum industry.[5] The diagram shows the bits of the network that are more or less visible to the public. How they communicate, strategize, and cooperate with one another to promote the industry is much less apparent. The petroleum sector itself is a complex set of relationships between exploration and production companies; consultancies specializing in geological/aerial surveying and analysis; environmental management consultancies; specialist drilling and fracking companies; engineering, site preparation, transport, and pipeline companies; and delivery firms like Z Energy and Caltex.

Petroleum industry promotion and networking with government, businesses, and supportive foundations is facilitated through *trade associations, lobbyists, conferences, and industry-focused media and front groups,* with PEPANZ at the forefront. PEPANZ also convenes the invitation-only Petroleum Club

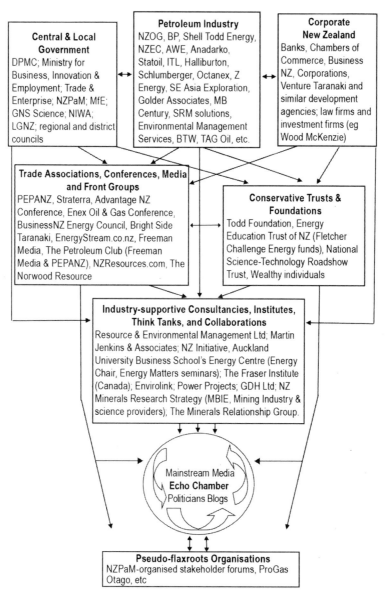

Figure 4.1.   Key Components of the Petroleum Industry Promotion Network in New Zealand.

Source: Adopted from R. E. Dunlop and A. M. McCright, 2011, "Organized Climate Change Denial," in *The Oxford Handbook of Climate Change and Society*, ed. J. S. Drysek, R. B. Norgaard, and D. Schlosberg (Oxford: Oxford University Press, 2011), 147.

which meets every few months in New Plymouth, the center of the country's oil and gas industry. PEPANZ collaborates with Straterra, the mining sector's mouthpiece organization, because of their shared interest in exploiting the country's mineral wealth. Several small-scale media companies like Freeman Media and NZResources have focused on the extractive sector. Freeman Media has helped organize several oil and gas conferences and the Petroleum Club. EnergyStream is an online industry news site hosted by Venture Taranaki with funding and technical assistance from government. A few "front groups" like Brightside Taranaki, an organization supported by businesses that service or benefit from the oil and gas industry, have been established to highlight the industry's economic and community contributions.

Less directly associated with the petroleum industry, from the public's perspective at least, are a number of supportive *PR consultancies, academic institutes, think tanks, and collaborations*. A scattering of small to medium-sized legal, technical, and environmental consultancies who provide services to exploration and production companies are occasionally asked to make media statements defending industry operations or serve as "independent experts" at public hearings, select committees, and local council briefings.

Conservative think tank The New Zealand Initiative has been outspoken about exploiting fossil fuels and minerals to boost the economy. A handful of academic institutes like the Auckland University Business School's Energy Centre maintain close relations with the oil and gas industry, business, and government receive industry funding and undertake research useful to the petroleum sector. In 2014 several universities submitted a joint proposal to government to fund a "Centre of Research Excellence in Earth Deformation, Resources, and Geophysical Exploration."

At the bottom of figure 4.1, oil and gas companies and PEPANZ are actively involved in what Dunlop and McCright call the "echo chamber" of *political discourse, mainstream media, social media, and the blogosphere*. We will examine later how the blogosphere has been used by lobbyists, PR consultants, and oil industry proponents to generate public support and undermine the credibility of opposition party policies and environmental groups.

Behind the scenes, with no apparent affiliation with the oil and gas industry, are what Dunlop and McCright termed "astroturf" (artificial grassroots) organizations. These claim to provide a neutral platform for discussion and debate and/or portray themselves as presenting "facts" about the benefits of oil and gas development while condemning environmental extremism. Either way, they are ultimately involved in industry advocacy and defense. Pro Oil and Gas Otago is a recent example of such a local petroleum advocacy group albeit with international connections. It was established in Dunedin by a small group of businessmen at the beginning of 2014 when Oil Free Otago's campaign against deep-sea exploration by Anadarko and Shell was grabbing the headlines.[6] The group claimed to be non-political, although two founding members were city councilors and the organization's spokesperson had strong links to the National Party. The purpose of the group, according to their spokesperson, was to support oil and gas development and counteract "scaremongering" by opponents of oil and gas exploration. She claimed that growing support for the group showed the "silent majority" favored exploration, and that petroleum industry people were "delighted by the public response."[7] After its launch, the group made contact with PEPANZ, petroleum industry lobbyists, and advocacy groups around New Zealand, and established communication with pro-fossil fuel groups overseas like the America's Natural Gas Alliance. There was talk of scheduling a pro-petroleum rally, meeting with oil companies and lobbying the Dunedin City Council to support petroleum exploration. When the organization launched a website in 2015, their promotional strategy shifted from attacking petroleum opponents to acknowledging climate change, promoting gas as a "transition fuel," and encouraging reasoned public debate.

The oil and gas industry in New Zealand has engaged in a range of activities to promote and defend itself, in many instances with the tacit knowledge and cooperation of the state. Four key strategies have been:

- PR spin and "reasoned" debate,
- Neutralizing environmental opposition,
- Influencing government policies and regulations, and
- Co-opting communities and "partnering" with iwi.

Given the presence of so many overseas-based oil companies in New Zealand, it is not surprising that some of these strategies have a familiar ring when compared with industry strategies elsewhere.

PR SPIN AND GENERATING "REASONED" DEBATE

This strategy has a dual purpose: capturing the support of what is colloquially known as Middle New Zealand while at the same time establishing a distinction in the popular consciousness between sensible, hard-working, outdoors-loving Kiwis and emotional, ill-informed, "tree-hugging" environmental extremists. The strategy relies on using purposefully slanted discourse, which Kidner[8] reminds us is a form of power-infused action, to construct such protest groups as dangerous, irrational outsiders.

PEPANZ, Conferences, and the Media

PEPANZ, the national mouthpiece for the upstream (i.e., exploration and production) oil and gas industry, has been a key institution for honing and implementing this particular strategy. Established in 1972, members include big international corporations as well as junior explorers and associate members who provide goods and services for the industry. PEPANZ's main advocacy activities are "working with local and central governments" to ensure the regulatory and commercial frameworks maximize returns and benefit "the community"; and "seeking to increase community and government understanding of the upstream petroleum industry." [9] PEPANZ claims that "growing the oil and gas industry in a responsible way will protect the environment and enrich communities."

In 2012, as we have seen, the National government moved to implement the New Zealand Energy Strategy, initiated a new block tender process, and increased its overseas marketing efforts to promote oil and gas exploration. Sensing this could be an important turning point for the industry, PEPANZ members agreed to increase staff numbers and the budget, and reorganize the governing board to include major explorers and producers with political clout. [10] New Zealander David Robinson was appointed CEO in early 2014. Prior to his appointment, he had worked for Shell New Zealand before he became Commercial General Manager for Z Energy, a company created when Shell sold its New Zealand distribution network. Robinson assumed office at a time when the industry needed to strengthen its working relations with the National government in undertaking legislative and regulatory reforms, improve public support for oil and gas expansion, and respond to the environmental movement's high-profile opposition to fracking and deep-sea exploration. [11]

Robinson early on adopted the tactic of engaging in fictitious public debate via the media, stating the case for the importance of the petroleum industry to the country and then pointing out how environmental groups and community anti-oil activists were misguided and ideologically blinkered. The problem with the manufacturing debate strategy is that in a liberal democracy, major decisions about the country's development path have to be based on more than public opinion polls and skillful repartee in the media. Ultimately they involve extensive evidence-gathering, formulating options, policy development, and reasoned judgments by the people's representatives about the pros and cons of proposed courses of action. The manufactured debate strategy attempts to skew those choices by circumventing objective evidence-gathering, cost/benefit analysis, and assessment of options that are part of normal policy development. In this instance the public and lawmakers are asked to trust assertions by the petroleum industry, based on selected "scientific" evidence and overseas operational experience, that the benefits of expanded oil and gas development far outweigh the costs. To paraphrase economist Geoff Bertram, beware the promoter who ignores or minimizes the costs and harmful spill-over effects of a proposed development or an industry's ongoing operations.[12] The risk with employing the manufactured "reasoned" debate strategy is that environmental critics and independent scientists may successfully counter industry claims with evidence of their own that could influence public opinion and policy-making. To manage that risk, the petroleum industry has been forced to adopt a parallel strategy of co-opting or neutralizing "moderate" environmental opposition (see below).

Robinson and PEPANZ consistently emphasized three propositions in their public debate messages and PR spin to establish trust and support for the industry:

1. Growing the petroleum industry will benefit the New Zealand economy and society;
2. An environmentally responsible industry in partnership with government and communities will mitigate the "less than minor" costs of petroleum development; and
3. A caring, forward-thinking industry is leading the transition to a low-carbon economy.

Let's briefly examine how these claims were injected into manufactured public debate. The first claim had to do with the economic and social benefits of petroleum development. From the time he became PEPANZ CEO, David Robinson made the economic and social benefits of an expanded oil industry one of the central themes of his public discourse. In an early media interview he pointed to the more than $1 billion in direct taxes and royalties the oil and gas industry paid at the time. The industry, he claimed, had the potential to give a major boost to the economy given the country's eighteen basins and the size of the New Zealand's exclusive economic zone.[13]

PEPANZ organized the inaugural New Zealand Petroleum Summit in 2012. In his media release, Robinson stressed the benefits that an enlightened, "world class" oil and gas industry could bring to the country and local communities. But he was also mindful of growing public concern over government's plans to significantly expand the industry:

We want to lead the world in innovative oil and gas technologies, safe and robust processes, energy security and skills development without compromising our environment or health and safety. Our communities must come on the journey with us.[14]

Robinson's rhetoric portrayed the industry as both "world leading" and as a caring home-grown industry concerned about "our" environment and "our" communities. The conference program featured a handful of local government representatives, a token environmentalist, and a Māori iwi engagement consultant who told oil folks the "appropriate" ways to get alongside tangata whenua (indigenous people). In spite of the emphasis on community and iwi engagement, the Summit was not open to the public. PEPANZ's Chairman Chris Bush opened the conference with a rousing address about the benefits of the petroleum industry to the New Zealand economy, the "real people" involved in the business, and the challenges that lay ahead of the sector as it grew.[15] Bush called on his audience to "take the time and explain [i.e., propagandize] why it is important to grow oil and gas exploration here in New Zealand so communities have an understanding of the benefits and the future energy situation we may face without it."[16]

Exploration activities by major international corporations onshore and offshore increased over the next couple of years. In 2014 the arrival of Anadarko's deep-sea exploration ship off the coast of Otago in the South Island sparked local protests and provided Robinson and other industry spokespersons with the opportunity to again argue the benefits of an expanded petroleum industry. He told the *Otago Daily Times* the exploration phase would bring relatively modest economic activity to the region initially, but if an oil or gas field went into production then there could be big spin offs for the wider community.[17] He claimed this was how a viable industry was created in Taranaki, although he side-stepped questions about the negative impacts.[18] Two years later, Robinson's replacement as PEPANZ CEO Cameron Madgwick went further, telling a Dunedin business gathering a significant petroleum discovery in one of the country's basins "would quite simply lead to an economic revolution!"[19]

The second debating claim was that oil and gas is a safe, environmentally responsible industry. As noted previously, greenwashing is the process whereby corporations appropriate environmental messages, portray themselves as caring stewards, and reinterpret environmental problems as conducive to technical solutions within a supposedly more responsible capitalist global economy. PEPANZ's Robinson proved adept at greenwash or, as linguists say, "re-contextualising" environmental values and goals as those shared by the petroleum industry. He told a *New Zealand Herald* reporter shortly after taking up his position as CEO, "I absolutely acknowledge the green movement for changing the way we view the world and how every industry operates. We [presuming to speak for Middle New Zealand] have a tremendous amount to thank them for and I think most of us share the ideology." He stated that he loved the outdoors and would welcome "a clean, green energy discovery to supplant hydrocarbons." But he cautioned we have to be "practical" (implying his opponents were impractical): oil and gas would continue to be the primary fuel for years to come.[20] Months later, Robinson strongly rejected claims by climate activist Bill McKibben during his visit to New Zealand that transition to a clean-energy economy was being delayed by a fossil-fuel industry "who used its wealth and political influence to prevent change."[21]

Speaking of greenwash, in his earlier address to the 2012 Petroleum Summit PEPANZ Chairman Chris Bush enthused about the petroleum sector's environmental credentials. His tone was positively evangelical as he sought to link public pride in the New Zealand's clean green image with the oil industry's responsible environmentalism: "Wow! Isn't that beautiful!" he exclaimed. "As a country we don't have to choose between growing the oil and gas industry and keeping our environment clean and pristine!"[22]

Both Bush and Robinson conveniently overlooked the growing number of petroleum incidents (accidents, spills, leaks) occurring offshore and onshore, which were required to be reported and investigated by the government's newly established High Hazards Unit. Between March 2012 and March 2015, approximately 150 such incidents or "petroleum dangerous occurrence notifications" were recorded.[23]

Between the Petroleum Summit in early 2012 and the Advantage New Zealand conference in October 2014, the Parliamentary Commissioner for the Environment (PCE) prepared and released two reports on fracking and government's regulatory regime for overseeing oil and gas development. In anticipation of the release of the Commissioner's 2012 interim report, PEPANZ issued a media statement providing a "factual" background on fracking in New Zealand. The organization also included Todd Energy's submission to the PCE's inquiry on its website. The submission not surprisingly concluded that fracking "as practiced in New Zealand" was safe.[24]

Cameron Madgwick replaced David Robinson as CEO in mid-2014 when Robinson resigned to become CEO of exploration and production junior New Zealand Energy Corporation. Madgwick had a legal and community engagement background. On PEPANZ'S website he declared his vision was to provide "solid and accessible" information to communities and grow the financial contribution the sector made to New Zealand's economy "in an environmentally responsible way." Echoing these themes, the program of the organization's Advantage NZ Petroleum Summit in October 2014 included a strong emphasis on the industry's environmental credentials. Unlike the first Summit two years earlier, this time a large crowd of environmental activists protested outside the venue.

The third claim in the manufacturing debate strategy was about the petroleum industry's important role in transitioning to a low-carbon economy.

This is a relatively recent theme picked up apparently from PR spin emanating from some of the larger petroleum transnationals overseas. PEPANZ's Robinson acknowledged on several earlier occasions that climate change was real, and although the world economy was dependent on petroleum products, the industry had an important role in transitioning to an alternative energy future. Internationally, some commentators suggest the oil and gas industry is betting on a long transition process while attempting to position natural gas as a transition fuel. PEPANZ Chairman Chris Bush seemed to adopt that perspective when discussing New Zealand's transition prospects during his 2012 Petroleum Summit address. He avoided mentioning climate change altogether, suggesting "energy" was the most important issue of our time. He maintained that while people might want to switch to renewables tomorrow, the "reality" was that renewables made up only 40 percent of New Zealand's energy supply (80 percent of electricity generation). Even if there was significant investment in renewables, he argued, the country would remain heavily dependent on oil and gas for a long time to come.[25]

In the lead up to the national election in September 2014, climate change and expanded oil and gas development were relegated to the periphery of public debate. The National Party was keen to keep the incongruity between the government's petroleum expansion plans and climate change policies off the agenda. The maneuver wasn't helped when news broke that the Rockefeller family was joining a coalition of philanthropists pledging to divest themselves of $US 60 billion in fossil fuel assets. Two days after he took over as CEO of PEPANZ, Cameron Madgwick was asked by Radio New Zealand for comment about the Rockefeller move. Madgwick stated that he "welcomed" the Rockefeller decision, noting that the petroleum industry was helping in the transition to a low-carbon economy. When the interviewer asked *specifically how* the New Zealand industry was planning for a low-carbon economy without petroleum, Madgwick was caught off guard and resorted to a fabrication which he evidently hoped (correctly as it turned out) wouldn't be challenged. "We're all invested as an industry, in partnership with communities and government, in developing a plan over the longer term to transition away from oil and gas."[26] Madgwick would have known that neither the industry nor the government were working on such a plan, and they certainly weren't "partnering" with communities to reduce emissions and dependency on oil and gas.

Industry-Supportive News Sites and Specialist Consultants

Besides company websites, news and information about the oil and gas industry has been channeled through a handful of small-scale media companies that have emerged in recent years. These companies have made their mark by establishing industry-focused subscription news sites and other on-line products, feeding stories to mainstream media (MSM), and promoting and managing industry seminars and conferences.

Freeman Media has been one of the more prominent of these information-brokers. They present themselves as "an experienced and savvy information services provider to the New Zealand corporate community,"[27] in particular the petroleum industry. Freeman has collaborated with PEPANZ and NZPaM in organizing the annual Advantage New Zealand conference and also helped run the quarterly Petroleum Club held in New Plymouth, the main town in Taranaki (see below). They also organize the annual New Zealand Downstream Strategic Forum, administer a website called "New Zealand Petroleum Exploration Opportunities," and publish newsletters *Inside Resources* and *Energy News*.

A relative newcomer is subscription-only NZResources.com, an online mining and petroleum news site. It is owned by Silver Budgie Pty Ltd, a Western Australian partnership between technical director Carl Knox-Robinson and editorial director Ross Louthean. Louthean previously ran three different publishing and conference groups in Western Australia. He frequently publishes informational articles and op ed pieces in mainstream media like the *Otago Daily Times* (*ODT*). The links between NZResources.com and the *ODT* are particularly close, though not widely known. The two main reporters for the NZResources are also the *ODT*'s senior business reporter Simon Hartley and business editor Dene Mackenzie. Hartley is regularly "comped" to attend and report on oil and gas events like PEPANZ's annual Advantage New Zealand conference.

The expanding oil and gas industry has also benefitted from advice and promotional support from specialist consultants, many of whom have crossover experience in the petroleum industry, government, law and investment, and the media. John Kidd of Woodward Partners, a share brokering and corporate advisory firm, is a case in point. Kidd is currently head of research. Previously he was head of Pacific research at Edison Investment Research, where he prepared the first annual New Zealand Petroleum Yearbook in 2012. Before that he had been a principal consultant with Shell International, a senior advisor with New Zealand Treasury, and then head of Research at McDouall Stuart investment advisors. While at McDouall Stuart, Kidd produced the 2009 *Stepping Up* report for then Energy Minister Gerry Brownlee. He is a paid-up member of the exclusive Petroleum Club, and was a guest speaker at the 2013 Petroleum Summit and the October 2014 Advantage New Zealand conference.

David Coull is a lawyer with Wellington firm Bell Gully specializing in commercial law particularly in the oil and gas sector. Coull is also on the

Board of PEPANZ. In 2015 he was named in *Who's Who Legal: Energy 2015* as one of twenty-four most highly regarded international energy lawyers. He is on the governing board of PEPANZ, representing his firm and other associate members. In addition to his legal advice, Coull produces occasional updates and commentaries on legislative developments and public issues that affect the New Zealand petroleum industry.[28] In 2013 Coull was invited to address the Advantage New Zealand conference. In his presentation he suggested the oil and gas industry in New Zealand was at a tipping point.[29] Although much-needed legislative reforms were underway, there was an upsurge of public concern, lobbying, and protest regarding climate change, fracking, and off-shore exploration. To deal with these challenges, Coull recommended that the industry concentrate on gaining a social license by engaging with "interested/ affected parties" such as iwi, landowners, and local councils.[30]

Well-known television journalist and Radio NZ host Linda Clark moved to law firm Kensington Swan in 2014, where she combined her legal training and media skills as council specializing in government relations, public policy, and regulatory issues. She also became co-leader of the company's crisis management team, providing advice on corporate risk to businesses including the petroleum industry.[31] Kensington Swan is a member of PEPANZ, as are ten other New Zealand and international law firms.

Risk management, particularly reputational risk, is an important field of petroleum industry consultancy. Chris Bush, PEPANZ's former chairman, runs a New Plymouth-based consultancy called SRM Solutions Ltd. SRM provides "practical strategy and risk management solutions to CEOs and Boards" no doubt based on Bush's experience starting up and/or leading three New Zealand petroleum exploration companies. Spiro Anastasiou is another crisis management specialist who was invited to speak at the 2013 Petroleum Summit. He has high-level experience managing government's public communications around various crises, including the Christchurch earthquake. After his government work, Anastasiou became a partner at SenateSHJ, a major international public relations, communications, and change management firm. SenateSHJ services, which extend to the petroleum industry, include corporate communication, social media, lobbying, regional and rural communication, and social marketing. The company is a member of PEPANZ, as are consultancies such as Minter Ellison Rudd Watts, Hale and Twomey, and BRG who include strategic advice, government advocacy, and PR among their services.

These industry-focused media services and specialist consultancies are in effect the sinews that help connect the oil and gas industry "promotional" and defense network in New Zealand. They facilitate communication among the main players and tailor messages to the wider public. They also provide knowledgeable "insider" advice to oil and gas companies on political developments, risk management, and corporate strategic planning.

Collaborating with Supportive Business Organizations and Think Tanks

In a small country like New Zealand networking among like-minded corporations, business associations, and a handful of conservative think tanks has been a useful way for the petroleum industry to gain allies and receive help in communicating its promotional messages. Straterra was formed in 2008 as the collective voice of the minerals and mining sector. It maintains reasonably close working relations with PEPANZ since their operational boundaries overlap. Straterra's membership includes two state-owned enterprises, Crown research institutes GNS Science and the National Institute for Water and Atmosphere (NIWA), along with five leading law firms and three major accountancy companies. The organization administers the invitation-only Mining Club, a cousin of PEPANZ's Petroleum Club. It also co-hosts the annual Minerals West Coast Forum in cooperation with the Minerals West Coast Trust and helps organize the annual conference of the NZ Branch of the Australian Institute of Mining and Metallurgy.

Business New Zealand has been one of the oil and gas industry's strongest supporters. The organization's newsletters regularly contain articles about the industry, energy security, and climate change. CEO Phil O'Reilly from time to time issues media statements on developments in the sector, including reform of the RMA.[32] The Major Companies Group within BusinessNZ, established to ensure "NZ's largest companies are heard in policy, business and economic debate" includes BP New Zealand, Chevron New Zealand, Methanex, New Zealand Oil and Gas, Refining New Zealand, Shell New Zealand, Todd Corp., and Z Energy. Another group, the Business New Zealand Energy Council (BEC), has been instrumental in presenting the industry's case to government and the public. Members include BP, Chevron, NZ Oil and Gas, Statoil, and Z Energy as well as Straterra. The BEC is allied with the World Energy Council (WEC), whose members include most of the world's largest petroleum companies. The BEC sponsors New Zealand participants in the WEC's Future Energy Leaders program, which is modeled on the Energy Ambassador program run by the US Petroleum Institute. Energy "leaders" or ambassadors are promoted as role models for high-flying careers in what is portrayed as a technically sophisticated global industry. They are also encouraged to use their personal networks, social media, and speaking engagement opportunities to spread positive messages about the petroleum industry.

There are twenty-nine Chambers of Commerce around the country with over 22,000 members. Nationally the Chamber seeks to influence the regulatory and economic environment in which businesses operate and advocate on behalf of members.[33] Regionally, most Chambers have been vocal in their support of possible oil and gas exploration. They have released media statements, organized visiting speakers on issues supportive of oil and gas development and provided input to debates in their local media as well as lobbying their parliamentary and local council representatives.

Most regions in the country have semi-autonomous economic development agencies for whom oil and gas exploration and production are at least a distant prospect. They are aware of, and in some cases have supported (e.g., Southland), government-sponsored aerial mapping of their region's geology. But most, like the Gisborne region's Activate Tairāwhiti, appear to have shelved the immediate prospect of any on the ground developments given the downturn in the global oil market.[34] Venture Taranaki (VT) is an exception, actively supporting and advocating for the industry because of its long-standing contribution to the Taranaki economy. VT regularly issues media releases and promotes the industry on its website. In 2010 the organization released a major study called *The Wealth beneath Our Feet* on the potential of the petroleum and minerals sectors for the economy of Taranaki. The report was considered such a PR success that they released an updated version in 2015 entitled *The Wealth beneath Our Feet: The Next Steps*. VT's CEO Stuart Trundle provided an update at the 2016 NZ Petroleum Conference, mentioning possible initiatives that could help the industry to continue to expand. The industry and government have pointed to the study as making a convincing case for oil and gas development in other parts of the country. However, the government is hardly a neutral commentator on the merits of VT's study. VT receives funding from government agency NZ Trade and Enterprise, some of which has been used to develop a website called EnergyStream.co.nz. NZTE provides ongoing news, information and government policy updates for the website, which touts itself as a communications hub for the oil and gas industry and its suppliers. It promotes the industry, disseminates information (some of it government-supplied), and posts news and notices of upcoming events. The site's administrators also monitor mainstream New Zealand media, publishing links to positive news stories about the industry and answering critical articles. For example, when the WWF's 2013 report criticizing government's oil subsidies made headlines, VT's site published a rebuttal commentary and provided links to online UK petroleum industry resources for companies and industry supporters to use in answering the WWF's claims.

The New Zealand Initiative is a research and policy think tank formed by the merger of the right-wing Business Roundtable and the New Zealand Institute. The organization claims to be non-partisan and independent thinking in contrast to government departments and universities. Their published briefs, papers, and reports lean toward a liberal free market, anti-big government, growth paradigm. On its website, the organization says its aim is to develop "policies that work for all New Zealanders."[35] In late 2014 and early 2015 research fellow Jason Krupp produced two think-pieces designed to provoke debate on the potential role of extractive industry in boosting regional development. The first report, released in December 2014, was titled "Poverty of Wealth: Why Minerals Need to Be Part of the Rural Economy" and claimed minerals mining had the potential to reverse a "cycle of decline" in several regions.[36] In his second report, "From Red Tape to Green Gold," Krupp took a swipe at the environmental movement and criticized government foot-dragging over RMA reforms.[37]

NEUTRALIZING ENVIRONMENTAL OPPOSITION

Neutralization strategy is not just about countering environmentalists' criticisms or managing protest actions so they don't unduly disrupt conferences or field operations. It's about addressing the reputational threat that high-profile environmental organizations, community anti-oil movements, and independent scientists pose through (a) documenting and highlighting the impacts of oil and gas development on human health, communities, and the environment; and (b) raising awareness of climate change as an ecological problem and a challenge to unbridled economic growth.[38] More and more people begin to question whether increased drilling and government's commitment to growing the fossil-fuel-dependent economy will ensure their own and their children's future well-being and preserve the environment. These are threats the industry overseas has taken seriously and gone to extreme lengths to combat.

As yet in New Zealand, there is no smoking gun to suggest orchestrated campaigns like those overseas or the subversive corporate efforts that occurred in the 1990s to discredit the anti-native forest logging movement on the country's West Coast.[39] That doesn't mean that some of the same tactics aren't being applied today in particular instances. Public rhetoric and managing protests in conjunction with the state are the two primary tactics the New Zealand petroleum industry seems to have employed against environmental activists. Public rhetoric by ministers, company PR spokespersons, and industry-friendly consultants takes the form of questioning environmentalists' motives, casting doubt on their emotional stability, challenging their legitimacy and loyalty, and disparaging their protest efforts. Environmental activists simply make "noise," are always negative, and resort to "scare tactics" and "hysteria" according to consultancy firm McDouall Stuart.[40] At the height of the fracking debate, Straterra's policy manager Bernie Napp published an opinion piece in the *New Zealand Herald* suggesting "armchair experts" should get their facts straight.[41] Arguments by opponents of fracking "conflate emotion and reason which is not unusual in public debate; that's human but it doesn't make it right." He accused critics of being ill-informed and not separating "fact from innuendo."[42] Napp went on to try to dispel a number of fracking myths with a few "facts." Most were either half-truths or simply wrong, as several of the twenty-nine commentators on his article pointed out.

In his opening address to the 2012 Petroleum Summit, PEPANZ Chairman Chris Bush[43] launched a blistering attack on opponents of oil and gas development, accusing them of "dehumanizing the industry," "misrepresentation of the data," spreading "mistruths for political gain" (presumably a reference to the Green Party), and subjecting the public to "scaremongering and emotive debate." To combat what he termed "propaganda from naysayers," Bush called on the industry to "put the facts in front of people," rebut the mistruths, and "build public confidence in the way the industry operates." Underscoring the importance of the industry's collaboration with the state, Bush emphasized that "public confidence [i.e., a supportive electorate] is what the government needs if it wants to turn its goal of increasing oil and gas exploration into a reality."

Following the 2013 Petroleum Summit, where similar concerns about environmental critics and anti-fracking protests were expressed, PEPANZ decided it was time to strengthen its own public propaganda capabilities. In January 2014 they advertised for a Resource Management and Community Affairs Manager.[44] The purpose of the role, according to the job description, was "to assist the CEO in representing the industry across New Zealand to build understanding of the industry, present factual information to rebut misinformation and scaremongering, explain why petroleum resources are being developed, detail the economic benefits, and discuss community and environmental impacts."

Midway through the year, Cameron Madgwick took over the PEPANZ helm. With his community relations background, he soon found himself asked to comment on various protest actions and critical media articles about deep-sea drilling and onshore fracking. When Greenpeace activists interrupted proceedings at the October 2014 Petroleum Summit, Madgwick assured journalists the protests had been expected (which rather begs the question why PEPANZ hadn't come up with a plan to prevent the disruption).[45] He maintained the protests arose out of "a lack of understanding about the industry. There are a lot of mistruths out there and they tend to engender fear. If you don't understand something, being fearful is a natural reaction."

PEPANZ he said was working hard to eliminate some of that fear through "education . . . getting the scientific facts out to people in a way they can understand."[46] Of course, that also meant having to rebut mounting international research evidence of the harm the unconventional oil and gas industry was doing (see chapter 2).

In a 2016 interview with the author,[47] Madgwick stressed the industry's line that it was people's democratic right to protest and have their views heard.

> While I don't share their views they're entitled to have them heard. If it's blockading the entrance to our conference or a social media campaign, that's fine. We, the industry, have absolutely no problem with that.

Madgwick said their only concern at public meetings and conferences was a health and safety issue. Unlike most established petroleum countries, he maintained that PEPANZ had little communication with the police regarding protests. Police did their own threat assessments and took whatever action they thought necessary. However, if protests disrupted exploration or production operations, that was another matter. The industry had a right under law to operate in New Zealand and expected those rights to be honored by opponents.

The problem for large international petroleum companies looking to operate in New Zealand is that protest groups are not easily dismissed as self-interested NIMBYs (not in my back yard) or emotional, misinformed fanatics. There is a long history of environmental activism. Established environmental organizations (of which there are many) have long track records defending the environment, have done their homework, are extensively networked, and most are linked to well-informed, well-organized international groups. Increasingly PR rhetoric or tactics employed by the industry are well-known and counter-measures are shared around the environmental movement. Sometimes, as in the case of Chevron who is partnering with Statoil to explore the Pegasus Basin off the East Coast, activists even come from overseas and make life difficult for such companies.[48]

Until now, PEPANZ and the petroleum companies have largely followed a strategy of talking up the industry's credentials and safety, generating and managing public debate, and attempting to discredit environmentalists. The focus on the one hand has been on winning over the hearts and minds of Middle New Zealand while tackling environmental "extremists" through vitriolic rhetoric and control of protests. PEPANZ CEO David Robinson reiterated the two-pronged strategy during his 2013 Advantage NZ conference presentation, suggesting recent survey findings (which he misrepresented) showed almost half of New Zealanders were supportive of the industry.[49] The industry, he claimed, was winning "the battle."

That confrontational approach, however, seemed about to change. The organizing committee for the 2015 Advantage New Zealand conference apparently thought industry and government participants would benefit from hearing from someone who had direct experience dealing with highly organized anti-petroleum activists and community protest groups. They invited Colorado Oil and Gas Association CEO Tisha Schuller to deliver a presentation based on recent developments regarding unconventional oil and gas in Colorado. The title of her presentation, "De-escalating the Fracking Wars: A Colorado Perspective," was the same theme she and her staff had used when traveling throughout Colorado talking to community groups and environmental organizations. For Schuller and the petroleum industry she served, the trick to winning the hearts and minds of the majority was to render polarized conflict and confrontation illegitimate: "We don't want to energize protest" she said. Schuller and COGA staff presented themselves as caring, trustworthy citizens who provided "factual information" and were open to engaging in face-to-face dialogue. They wasted little time on extreme elements who couldn't be "talked around" to the industry's point of view. Schuller emphasized, "We want the 80% in the middle." Following her address, several panelists representing New Zealand oil companies discussed the need for "dealing with the growing voice of dissension of *hardened environmentalists*, committed solely to no drilling or extraction of oil or gas, and broaching little or no discussion on moving to non-fossil, renewable energy sources." [50] The tactics were typical of what some critics have termed "subversion" of organized anti-fracking and environmental movements (see chapter 1).

With the benefit of hindsight, it appears that Schuller's address and the subsequent panel discussion signaled the beginning of a shift in the New Zealand oil industry's promotional strategy *away from* confrontational polemics like those expressed in the 2012 conference toward "genuine" community engagement and iwi consultation. They were obviously picking up similar signals from National government ministers and NZPaM officials, as we noted in the previous chapter. Following the 2015 conference, several industry executives commented in the media about the need for open discussion and engagement with established environmental NGOs and communities, particularly outside Taranaki.

The flip side of this new approach was that the petroleum sector appears to have given up on attempts to sway the opinions of "hardened" environmentalists and climate change radicals. Acknowledging that oil companies and the government were seen as having vested interests and private agendas, one 2015 conference observer noted that "speakers and some delegates identified 'Middle New Zealand' or 'the other 80%' for future engagement." [51]

The consequence was a sharper demarcation in PR rhetoric and engagement tactics between moderate environmental groups, who could be included in public discussion, consultations and stakeholder forums, and "extremists" who would not listen or refused to engage. They could largely be ignored unless their protests turned to civil disobedience or violence.

In a 2016 interview,[52] PEPANZ's Madgwick confirmed impressions that the industry had shifted its public messaging and engagement approach, targeting concerned citizens and moderate environmentalists who were willing to engage and wasting little time on hard-core activists:

> Well, look, if Greenpeace or Stop Statoil or Climate Justice Taranaki or you name it . . . if there was an invitation to come and speak, we would accept that. We have gone out of our way to invite them to participate in our forums. And they choose not to, and I understand why they choose not to. And that's fine. The question is "are you going to put your investment in communications into somebody that doesn't want to listen to you?" No, of course you're not.

It is not entirely clear from interviews, media statements, and conference discussions whether Madgwick and petroleum industry executives actually do understand why such committed environmental organizations choose not to engage in industry forums or consultative hui. At least some experienced environmental campaigners feel that to participate in gatherings where the industry (with or without state involvement) sets the agenda and controls the process is to play the disempowerment game.[53] It allows the industry to claim how open and reasonable it is, and reduces any evidence-based campaign of struggle to a vacuous debate.

Utilizing "Independent" Experts and Academic Institutions for Legitimacy

As noted earlier, numerous international examples have emerged in which scientists in the private sector and academia have acted as apologists for the oil and gas industry. They make "expert" statements in the media, serve on industry boards, and are contracted to testify for the industry in public hearings. Sometimes these relationships are covered up, in which case their behavior is unethical and their work suspect. In New Zealand there have been a few instances where oil companies, government agencies, or local councils have contracted with academics or consultants to produce reports that provide an aura of legitimacy to the company's views or consent applications because of their perceived independence. Examples include Gisborne Council's contracting of Lincoln University researchers to undertake a literature review on fracking, Taranaki Regional Council's review of fracking with consultant input, and the Ministry for Primary Industry's much-criticized 2014 study of landfarm contamination involving dumping and recycling of drilling waste.[54] Even ex-Minister for Conservation Dr. Nick Smith, prior to his reinstatement as Minister for the Environment, was cajoled into putting his name to an article ghost-written by PEPANZ staff and appearing in their online newsletter criticizing of fracking "hysteria" sweeping the country.[55]

There are as yet no known examples in which academics or institutions have been covertly paid to support the oil and gas industry with their research, publications, or public endorsements. There are, however, instances where indirect industry influence may have been brought to bear, or perceived to have occurred, through funding support, contracted research, or research project collaboration. Examples are GNS Science, a Crown research institute who are often contracted by the petroleum industry to undertake research and produce geo-survey data. GNS claims its Energy and Resources research program is about "Sustainable energy development and resource use for wealth creation."[56] The organization has been paid to provide speakers at various mining and petroleum conferences, and has been asked to make expert statements to the media and submissions to government regarding the importance of the mining and petroleum sectors. The government, as noted earlier, has awarded GNS millions of dollars in research grants to undertake surveys, mapping, and data-gathering on behalf of the oil and gas industry.[57] GNS Science in turn is a sponsor of the Petroleum Club and the annual Petroleum Summit, runs "education" programs for iwi and hapū (kinship group, subtribe) about oil and gas exploration, has staff specializing in community engagement and Māori consultation techniques, and its scientists frequently champion and defend the industry in the media while claiming to be independent and neutral.[58] For example, in 2015 GNS scientist Chris Hollis speaking on Māori Television's *Te Kaea* program stated, "New Zealand needs to take responsibility for the resources it uses, and so we have local oil and gas to explore for and to use locally—we should take that responsibility rather than relying on other places in the world to provide us with that resource."[59] Weaning the country off petroleum dependency was not something he was prepared to discuss.

Another example is Auckland University Business School's Energy Centre and Chair in Energy and Resources, both of which are endowed by the Energy Education Trust of New Zealand. The Trust was established with funding from Fletcher Energy Limited after it was sold to Shell. The Trust's stated purpose is the advancement of tertiary education and research in New Zealand in disciplines relevant to the oil and gas industries. The Centre hosts an annual seminar series for academics and the energy sector called "Energy Matters." It also provides a number of postgraduate scholarships, postdoctoral fellowships, and honorary/visiting fellowships for researchers and people in the oil and gas industry,[60] all above board, albeit reflecting a clear commitment to close working relations with the industry. However, such cordial arrangements can provide a platform for "expert" advocacy on behalf of the industry and challenging the credibility of industry-threatening research. For instance, shortly after the Royal Society of New Zealand released a report in April 2016 setting out steps to a "thriving low-carbon economy,"[61] Auckland Energy Centre honorary fellow Frank Duffield waded in with an opinion piece in the *New Zealand Herald*[62] warning of the costs of transitioning too rapidly to renewables and arguing that natural gas was an important transition fuel. It was a theme that was popular with PEPANZ and the international petroleum industry in the climate debate. Duffield, it turned out, was a partner in a natural gas shipping business.

Using Websites, Social Media, and Influencing MSM

PEPANZ hosts an interactive website set up several years ago to help get its messages out about the industry. Recently, in addition to the standard FAQ section, PEPANZ has developed several websites linked to their home page dealing with topics of significant public concern. According to CEO Cameron Madgwick, the impetus for this move came after a 2013 public opinion survey the organization funded showed (according to the consultants at least) that most people were open to hearing about the industry but the industry's messages weren't getting through. Two of PEPANZ's supplementary websites focus on acoustic imaging surveys (or seismic surveys) and deep-sea drilling. A third website on climate change and the transition role of natural gas is in preparation. As Madgwick explains:

> We just thought it was appropriate to have the facts out there so we developed a website around [each issue] and there's a booklet that goes with each one. We examine the activities and some of the criticisms that have been raised. . . . So you know, we don't shy away from it, but we want to say "Well, look, this is what we think. This is why we think this, and here's the scientific evidence." You know, we won't say anything unless we can validate it. We've taken quite a while to work out how we think . . . we may be wrong . . . but how we think *the public would like to access that information.*[63]

These websites have been developed to also respond to publicity campaigns and ongoing protest action by environmental groups and Māori activists. The websites are therefore a useful tool by which the industry is able not only to generate "reasoned" debate, but to attempt to rebut the contrasting research evidence and warnings of risk environmentalists have compiled from their own investigations and international communication networks.

The New Zealand oil and gas industry has begun to pay attention to the role that social media can play in spreading positive messages about the industry's benefits, safety, and environmental responsibility. PEPANZ uses YouTube to disseminate informational videos about the industry, and according to Cameron Madgwick is considering how it can utilize other forms of social media. Venture Taranaki spreads promotional messages via Facebook, Twitter, and YouTube as do a number of the larger oil companies. Some companies have staff or PR consultants with specialist communications and social media skills but nothing on the scale and level of sophistication of larger corporations in North America and Europe. PEPANZ's previous CEO David Robinson early in his tenure argued that the industry needed to do a better job of communicating to the general public, and one way of doing that was through social media. Several speakers at the 2014 Petroleum Summit echoed a similar message. Haliburton Australia's David Guglielmo suggested the mistruths and controversy surrounding fracking had arisen in part because environmental groups were much better at communicating through social media. The industry in Australasia he said was a bit "old school." [64]

Nevertheless there is at least circumstantial evidence that oil industry employees, lobby groups, PR consultants, and pro-oil advocates have engaged in the occasional ad hoc social media campaign to manufacture debate. Most cases have involved individuals or cohorts of individuals (e.g., New Zealand-based shareholders) rebutting or disparaging industry critics in the comments sections of online news stories, op ed pieces or blog sites. Such attacks are at least implicitly aimed at influencing public opinion and capturing the attention of mainstream media.

In August 2013, for instance, the Aotearoa Fractivist Network reported a sudden rash of posts on their Facebook page in response to critical comments about Anadarko's drilling ship arriving in the Marlborough Sounds between the North and South Islands to begin exploration. Many of the posts strongly defended the industry and ridiculed critics' statements. When the Facebook page administrator investigated, it turned out one frequent contributor was a retired British migrant living in Blenheim who had previously worked for Texaco. Another individual attempted to infiltrate a related Facebook group titled "John Key Has Let the Country Down" and dispense "factual" information to counter anti-government and oil industry comments. Investigations revealed the contributor was a woman who worked for a petroleum PR consulting company in Taranaki. Other commenters on the same Facebook site who consistently defended government policy and the petroleum industry turned out to be staff members of two state-owned enterprises. [65]

Some campaigns have been more blatant. In November 2013 Greenpeace organized protests against deep sea exploration on the west coast of the North Island. According to a TV1 nationwide opinion poll, 81 percent of the public were opposed to deep sea drilling and supported the protests. Four days later, Yahoo New Zealand online news ran an unscientific opinion poll asking "Do you support oil drilling off New Zealand's coast?" The poll was accompanied by a stock photo of an operational oil rig off the coast of Taranaki. During the first few hours of the poll the numbers for and against drilling ran approximately even. After word of the poll circulated, by mid-afternoon 67 percent of responses *supported* offshore drilling. Allowing for possible bias in the way the question was stated and inclusion of the oil rig photo, either there had been a massive shift in public opinion in the four days since TV1's survey, which seems unlikely, or the poll had been "blitzed" by pro-oil supporters. However the poll was manipulated, many readers would have come away with the impression that it represented national public opinion.

Meet the Flakkers

Behind the world of corporate communications, mainstream media, and the blogosphere there is an informal, covert domain of online activity populated by an assortment of opinionated individuals and committed petroleum proponents which, borrowing from Herman and Chomsky,[66] we might call the *flakkers*. They pop up anonymously in MSM comments sections or social media championing the oil and gas industry and challenging the messages and legitimacy of the industry's opponents. As Hager noted,[67] their aim is not only to answer petroleum industry critics but influence which stories are covered in the mass media and how the issues are shaped. These days, mainstream journalists pay considerable attention to what is "hot" in the social media.

One of these below-the-radar provocateurs goes by the patriotic-sounding moniker "Kiwi." Tracking some of his online communications reveals that he is a male based in New Zealand and closely associated with the oil and gas industry. He has access to detailed insider information and maintains contacts with people in PEPANZ, Straterra, and various petroleum companies. He is knowledgeable about the petroleum industry locally and internationally, and has access to a wide range of global studies, reports, technical documents, media releases, conference proceedings, and even industry insider information from which he presents authoritative-sounding "facts" and answers industry critics. He shows up regularly in online comments sections attached to MSM stories and on Facebook pages. He appears to be systematically monitoring and responding to online MSM stories, editorials, opinion pieces, and letters to the editor critical of the petroleum industry. He has time to spend online commenting and rebutting critics of the industry, so his activities are either part of his job description or he's not in the workforce. Whether by coincidence or through personal contacts, Kiwi occasionally adopts some of the tactics used by overseas petroleum industry PR experts.[68]

For example, during 2013 several news items appeared in the regional newspaper, *Hawke's Bay Today*, about oil and gas development and fracking. These stories attracted several critical online letters to the editor. Kiwi systematically responded to these anti-petroleum letters as they appeared. His responses consistently took a pro-industry line, while presenting himself as an objective, detached but informed observer concerned that the public learn the "truth."

In mid-October 2013 following TV3's news host John Campbell's now famous confrontation with Minister of Energy and Resources Simon Bridges, a story appeared on TV3's website about mounting concerns over Anadarko drilling in the Taranaki Basin.[69] The story revealed that Anadarko's own environmental report admitted there was a 50 percent chance that oil from any drilling accident would wash ashore before it could be contained. The story remained on TV3's website with a section inviting public comments. Over a two-day period, ninety-one comments and replies to comments were received, of which Kiwi contributed nineteen, or 21 percent. Here are some unedited samples:

Kiwi: If this was Norway everyone would be cheering at all the oil wealth coming their way.

Reply: [Norway has] a 78% tax on private oil companies. Definitely aint gonna happen in NZ with John Key at the helm, that's too bigger cut from his rich cronies pockets. That's what I call a government who handles resource wealth properly, note the incoming Conservative party, tax cuts & increased spending. Norways wealth is about to go down the gurgler just like the NZ economy is now.

Kiwi: The reason why Norway has a high tax rate is because they have already found the oil. NZ still has to find the oil, thats why the tax rate is lower.

Reply: In Norway it [the petroleum industry] is run by a state owned company, because there the government acts in the best interests of the people and not international corporations.

Kiwi: Are you saying that we should become like Norway and have lots of oil wells?

Reply: Are you Simon Bridges? You sound willfully ignorant enough to be him.

* * *

Kiwi: Is TV3 going to talk about all the advancements in capping stacks [technology to control oil well blow-outs] that have taken place in the last few years? I think it would add to everyone's understanding and might make people feel more comfortable with deep sea oil wells.

Reply: Kiwi I get the feeling you are an industry social media spin Dr or Damage Control agent . . . or perhaps Simon Bridges himself . . . given that you seem to be so willing to overlook both the ministers absolute incompetence and the massive and proven risks to this country. Your online name "Kiwi" reeks of irony as your willingness to see this country put at long term risk for some temporary financial gains, is far from patriotic.

        You "Kiwi" need to take a trip around the New Zealand coastline, soak it all up and enjoy it while it remains in relatively good shape. Then let us know if any percentage of risk of losing it all is worth a few million dollars easily earnt and easily lost. . . .

Kiwi: NZ is not a wealthy country. We have a huge current account deficit. We as a Country need the money. Your feelings are lying to you. I'm not a spin doctor I'm telling the truth. The meaning of life is not found in nature [!]. Keep searching.

Around the time of these exchanges, the *Gisborne Herald* published an op-ed piece which I submitted, suggesting on the basis of my research that local councils and communities should take a hard look at the risks of developing oil and gas as well as supposed benefits.[70] I mentioned as a case in point the issue of drilling waste disposal. Kiwi had apparently been monitoring the newspaper and submitted an online rebuttal the next day. First he maintained, presumably in reference to my piece, that decisions should be based on "science" and "facts," otherwise "emotional reasoning" and "fear" took over. Then he proceeded to list several ingredients contained in fracking fluids which were supposedly used by households every day or in food production. Kiwi attributed the information strangely to Straterra, not PEPANZ. The list of ingredients repeated almost *precisely* a list in the presentation by the regulatory affairs manager of a major international corporation at the 2011 Houston petroleum PR conference (cited earlier). It will be recalled PR executives at the Houston conference were encouraged to use the list to convince the public fracking chemicals were harmless. Neither Straterra nor PEPANZ discuss drilling waste or list fracking chemicals on their website, so Kiwi would need to have had someone at Straterra obtain this information on his behalf. Unless of course Kiwi was himself a staff member of Straterra or PEPANZ, which will probably never be known.

Investigative journalist Nicky Hager introduced many New Zealanders to the world of intrigue that is the country's nascent blogosphere with his book *Dirty Politics*.[71] One of the leading characters was right-wing blogger Cameron Slater who edits *Whale Oil* blog. Slater had little time for the New Zealand Greens party (whom he labeled the "Green Taliban") and was scathing in his comments about radical environmentalists. At the time Hager published his exposé the oil and gas industry was not one of Slater's main causes. Or more accurately, he does not appear to have been approached very often by petroleum industry PR staff or consultants to write or run stories favorable to the industry. All that could change if there were a major expansion in the oil and gas industry, and an accompanying upsurge in environmental protests.

Nevertheless between 2012 and 2014 during a period of frequent protests against fracking and deep sea drilling, Slater published several pieces critical of the protesters and supportive of the petroleum industry. Some of these pieces were clearly Slater's work, but one was written by ex-Local Government Minister Rodney Hide criticizing Dunedin City Council's decision not to invest in fossil fuels.[72] Others gave the impression they had been written by an industry insider with intimate technical and statistical knowledge of the oil and gas industry, but published under Slater's name. The anonymity gave the author(s) a chance to present "facts" while adopting Slater's own down-and-dirty style of writing. A case in point was a Whale Oil posting referring to a media story announcing that two new exploration licenses had been awarded on the East Coast.[73] Greenies, the anonymous writer of the article said, would cry "fracking hell" at the prospect of all the new jobs on offer. The announcement would trigger more attacks on fracking by "lunatic conspiracy theory Greens." The writer highlighted a recent UK Royal Society report "declaring fracking was safe" if it was strictly regulated. The UK report hardly rated a mention in MSM throughout the country, but coincidently (or perhaps not) it was referenced in two public presentations by PEPANZ's David Robinson around the time of the Whale Oil blog posting.

INFLUENCING GOVERNMENT POLICY
AND REGULATIONS

Like other oil-producing countries, PEPANZ and petroleum companies operating in New Zealand have an interest in ensuring the government pursues policies and regulations that are conducive to profitable business operations. The main tactics adopted for influencing the government in this country include lobbying, patronage, drafting industry-friendly legislation and regulations, and revolving door arrangements.

Lobbying and Mateship

Lobbying in New Zealand is far from the large professional industry it is in more established petroleum-producing countries. Most exploration and production companies have someone on their staff whose brief includes stakeholder/government relations. A few people with oil industry experience have established private consultancies that undertake government relations work for oil and gas companies. And some law firms and PR companies have specialist staff who do compliance and advocacy work, as well as providing advice on presentations at resource consent hearings and parliamentary select committees. The oil and gas industry regularly lobbies to block or influence legislation that threatens its interests. The Minister for the Environment Nick Smith proposed a legislative framework for marine protected areas at the beginning of 2016 which would only cover the Territorial Sea (12 nautical miles) and not the Exclusive Economic Zone out to 200 nautical miles. Environmental groups expressed disappointment that 95 percent of the country's marine environment would be excluded after the Minister had earlier indicated the EEZ would be included. The NZ Greens party and World Wildlife Fund claimed the limited framework was the result of intensive lobbying by the oil and gas industry to keep the EEZ free for exploration, contrary to sustainable ocean management.[74]

Much of the day-to-day lobbying for the oil and gas industry is handled or coordinated by the PEPANZ. The lead person for policy development and government submissions is Policy Manager Andrew Saunders, who joined PEPANZ in early 2015. Saunders's role includes engaging with ministers, appearing before select committees and liaising with key stakeholders. From time to time members of the board are also called upon to support particular submissions or form a delegation to meet with ministers.

Saunders's counterpart at Straterra is Policy Manager Bernie Napp. Like Saunders, Napp has a policy background and a personal network of government officials. His role is to influence policy development and lobby for sector interests. In April 2014, for example, in the lead-up to the national election Napp and CEO Chris Baker presented a "Minerals Briefing Paper" to a gathering of Parliamentary MPs.[75] The paper included twenty-four recommendations for improving New Zealand's investment attractiveness to mining and petroleum companies. Following Environment Minister Nick Smith's January 2015 sweeping RMA reform speech, Straterra was one of the first lobby groups to respond, announcing they were establishing an extractive sector working group to "participate in the debate for streamlining the RMA."

Although there are 121 MPs, 18,370 public servants and numerous trade associations and consultants in Wellington, the capital tends to operate like a small town. It is loosely organized into broad policy sectors woven together by informal networks of regular work contacts and "mates"—bureaucrats, industry lobbyists, corporate staff, ministers, and executives who maintain contact via phone, email, coffee breaks, and lunch chats. For long-serving and/or senior staff, old boy and old girl networks are still relevant for making contacts and communicating across the public and private sectors. Frequent seminars, meetings, and conferences in the capital provide opportunities to network and hear presentations and debates on the latest policy issues. They also give ministers a chance to make important policy announcements to a mostly friendly audience, and private-sector representatives have an opportunity to express their concerns and discuss recommendations with politicians and government officials.

Some events are by invitation only or limited to an exclusive membership. PEPANZ and Freeman Media co-host the quarterly New Zealand Petroleum Club (NZPC) which meets in New Plymouth. The NZPC is "a forum for upstream oil and gas operators in New Zealand. Operators, explorers, service companies, rig owners and key infrastructure providers use the forum to discuss upcoming activities and share information for mutual benefit and the advancement of the industry in New Zealand."[76] Meeting attendance is restricted to paid members and invited guests, and the media are excluded. Information from the meetings cannot be distributed externally (though it is posted on NZPC's website), and the identity or affiliation of speaker(s) and participants is not supposed to be disclosed. The NZPC has the appearance of a private industry confab, but that is not its underlying function. It provides a regular opportunity for oil and gas executives, technical experts, central and local government officials and occasionally ministers to network and update one another on recent policy developments and industry issues. Quarterly meetings are sponsored events, and not merely by oil companies. State-owned enterprises GNS Science and NIWA have both been sponsors and are members of the club. Updates from PEPANZ and NZPaM are a regular feature of meetings. In August 2014 for instance an update on environmental regulations in the Taranaki region was provided by Liam Hodgetts, Group Manager Strategy for the New Plymouth District Council and Fred McLay, Director Resource Management at the Taranaki Regional Council (and ex-Shell Oil employee). Guest speaker at the evening dinner was MP David Shearer, the Labour Party's spokesperson on energy and resources at the time.

Patronage

Parliamentary MPs are rarely caught accepting outright bribes, but they know their constituencies and the sectors where their votes come from. Occasionally someone is accused of being unduly influenced by receiving financial support or other perks from a private donor. Labour's economic development and Māori affairs spokesperson Shane Jones gained a reputation for his consistently outspoken support for the petroleum and mining sector.[77] In late 2013 Jones entered the contest for leadership of the Labour Party, which he subsequently lost. A few months later, in April 2014, the New Zealand Greens accused Jones of accepting a donation toward his failed Labour leadership bid from New Zealand Oil and Gas company board chairman Roger Finlay, who was closely associated with the National Party. Jones claimed the donation had "not influenced or affected anything that I've been saying about the importance of the extractive sector."[78] Two months later he announced his resignation as a Labour MP to become the National government's economic development ambassador to the Pacific.

Helping Draft Industry-Friendly Policies and Regulations

In its efforts to develop policy and legitimate claims of a "world class" regulatory regime, the government found it useful to tap the available expertise of the oil and gas industry particularly in the early days when it was developing its oil and gas policies and establishing NZPaM. This in turn provided industry representatives with the opportunity to establish close working relations with officials and influence how regulations were drafted and implemented.

We noted earlier, for example, how John Kidd, then head of research at McDouall Stuart, wrote the 2009 *Stepping Up* report that led to the development of the government's first Petroleum Action Plan. Ex-PEPANZ Chairman Chris Bush provided advice on developing the government's health and safety regulations, based on his expertise and involvement in the industry-backed Be Safe Taranaki Trust. And when TAG Oil launched their East Coast exploratory drilling program in 2013, one of two international staff the company recruited for its East Coast office had experience preparing government environmental manuals and regulations on unconventional oil and gas operations. This person was also instrumental in preparing a 2012 report for the New Zealand Ministry of Economic Development and GNS Science about onshore unconventional exploration in New Zealand.

Many of PEPANZ's submissions on proposed policies, legislation, and regulations, take the form of "technical" additions or modifications based on industry experience and operational requirements. The organization has a staff member who specializes in nothing but this kind of work.[79] Some of these innocuous-appearing technical submissions have to do with potentially controversial issues. For instance, PEPANZ made a submission on the Resource Management Legislation Bill 2015 suggesting that to avoid "duplication," local councils should cede authority over hazardous aspects of resource consents to WorkSafe NZ who has legislative responsibility in this area. PEPANZ made a similar submission was made to the South Taranaki District Council's planning review. Environmental groups like Climate Justice Taranaki pointed out this would have the effect of depriving citizens affected by a proposed hazardous development of expressing their views and presenting evidence.

In his speech to the 2013 Advantage NZ conference Minister of Energy and Resources Minister Simon Bridges acknowledged the contribution the industry had made in helping develop legislation and draft regulations:

> Many of you have actively engaged with the Government on the review of the Crown Minerals Act, the introduction of environmental legislation into the Exclusive Economic Zone and on new health and safety regulations for petroleum operations. I thank you for your tireless efforts to help make our laws and regulations world-class.[80]

Revolving Door Arrangements

Although the National government's campaign to expand petroleum development is relatively recent, there has been considerable mobility between the industry and public sectors. This is to be expected in a small country with (at least initially) a limited pool of government officials with experience in the petroleum industry. The petroleum industry also offers financial and career opportunities for bureaucrats and trade association staff.

Critics of the New Zealand oil and gas industry have pointed to the cozy relationships that sometimes exist between corporations and regulators, characterized as "revolving door arrangements." Aside from the odd secondment, most of this movement of personnel back and forth between the petroleum sector and government is opportunistic. Nevertheless, an individual's experience in one sector undoubtedly benefits the other through a transfer of expertise, contacts, and "insider" information. This cannot help but influence the individual's perspective and decision-making, sometimes improperly so.

NZPaM is a case in point. It has hired a number of staff with international oil industry experience. The organization's previous director of petroleum, Kevin Rollens, was an Oklahoma oil man before joining the agency in 2010. In June 2014 he accepted a position as general manager of Singapore-based Mont D'Or Petroleum, with exploration and production operations mostly in Indonesia. Prior to Rollens leaving, Mont D'Or had successfully bid for a permit to explore a block of land on New Zealand's East Coast. After Rollens moved to Mont D'Or, he recruited Māori Pieri Munro who had been NZ-PaM's Manager Iwi Relationships. Munro was made Mont D'Or's New Zealand country manager, not least because the East Coast region was 43 percent

Māori and his home district. The same year that Rollens left, Deputy Secretary David Binnie resigned to work for Z Energy. A year later in 2015, Josh Adams assumed the role of NZPaM's National Manager Petroleum after a career in mining, working for Shell Todd Energy Services and Santos Oil and Gas (Australia), and as lobbyist for the petroleum industry in New Zealand. Adams helped write Todd Energy's briefing paper on fracking which was submitted to the Parliamentary Commissioner for the Environment during her 2012 review, and was subsequently published on PEPANZ's website.[81]

PEPANZ is another example of the revolving door. David Robinson had a background in the oil industry before he became CEO in December 2012. After eighteen months, he was recruited by Canadian-based New Zealand Energy Corporation (NZEC) to join the board of directors and become chief executive officer of the company's New Zealand operations. The following year in January 2015, Andrew Saunders joined PEPANZ as policy manager. Saunders had a background in government policy work particularly with the Ministry of Economic Development before it morphed into the Ministry for Business Innovation and Employment. Eight months later, Kevin Sinnot took up a position with PEPANZ as the organization's communications manager. He was previously a communications advisor/project manager for New Zealand Police, and earlier in his career had worked as a television and radio journalist in New Zealand and overseas. Sinnot replaced Deborah Mahuta-Coyle, who left after three and a half years to take up a position with Busby Ramshaw Grice PR consultants. The firm frequently does work for the petroleum industry. Mahuta-Coyle had previously been press secretary for the Labour Leader of the opposition and before that the Minister of Māori Affairs.

We noted in the previous chapter several central government officials who had a background in the oil and gas industry, or who left to take up positions in the industry. The Taranaki Regional Council's Director of Resource Management Fred McLay, a prominent example of the revolving door, has been at the center of claims of regulatory bias.[82] McLay left the TRC to work for Shell Todd Oil Services for two years developing wells to meet TRC's resource consents, before returning to TRC in 2003. There are similar instances among industry consultants, lawyers, and investment advisors.

CO-OPTING COMMUNITIES AND "PARTNERING" WITH IWI

Community Engagement

As noted previously, some exploration and production companies have made sporadic efforts at community/Māori engagement and at least paid lip service to earning a social license to operate. In some localities this has involved little more than providing charitable contributions, giving the occasional civic-minded speech to the Rotary Club, attending a hui, and maintaining a flow of sponsorships and similar good-neighbor gestures. But increasingly, expanding unconventional oil and gas operations have provoked backlash and even protest action from affected residents, forcing oil companies and trade associations to take more seriously the need for well thought-out community engagement strategies backed by a specific budget for the purpose. The importance of community engagement was highlighted during protests at the 2014 Advantage New Zealand conference, and discussed seriously at the 2015 conference. Its relevance is underscored by a national culture that at least in law if not always in practice acknowledges the Treaty of Waitangi, values the natural environment, prizes volunteerism and civic participation, and has a history of popular environmental "uprisings" when those values are threatened. People for the most part take consultation and public involvement in deliberations about key issues seriously.

At the same time, under current government policy settings, community and iwi consultation has little substantive influence on oil and gas development. The main purpose, when the industry talks about community/iwi engagement, is not to negotiate the scope and terms of a proposed development or consider whether it should proceed at all. Rather it tends to be about public relations spin, identifying supporters, defusing protests, and co-opting local leaders. STOS, for example, has devoted considerable staff time and resources on building "partnership" relations with the rural community of Tikorangi in the Taranaki region. While the company engaged in these consultation exercises it continued to expand its drilling operations, provoking protest action that received national media coverage. To contain the bad press and growing local resistance, the company began publishing a newsletter called *Taranaki Community Update*, held a series of community meetings, and established a Tikorangi Community Forum made up of community members and company representatives. The forum's agenda was however limited to discussing issues such as landscaping around the company's well sites and management of truck traffic.

New Zealand Oil and Gas (NZOG) established "community panels" in their main areas of activity in South Taranaki and the South Island. Representatives from business, youth, environmental, social services, and tangata whenua were invited to attend. The company claimed the panels "provide us with insight and guidance about community expectations, and help us understand local views and how we can meet community expectations."[83] NZOG said the panels also helped guide their "social investment" program (local charitable donations), which relates to the following tactic.

Buying Community Support

Some petroleum companies quietly acknowledge that their community and iwi "social investment" activities are appropriate compensation for the impacts of oil and gas operations on communities. But being seen to be making such investments is also useful for offsetting negative public perceptions of the industry, and responding to growing public concerns over global warming. To date there have been few reported instances of blatant community "bribes" similar to the Canadian pipeline company incident mentioned earlier. Most community funding takes the form of charitable, or at least charitable-appearing, donations or sponsorships. While companies naturally calculate the net returns from such "investments" for their bottom line, they strive to present themselves as civic-minded "local" companies with the interests of the community and nation at heart. As PEPANZ's Cameron Madgwick explains:

> Our industry are quite big social investors. You'd only need to travel to Taranaki to see the impact our industry has made. I think the way the industry would think about it is that they are a part of that community, and being part of that community means playing your role. And one of the things you can bring as an industry is money and time. But you find a lot of things the industry does, they don't want them spread on the front page but they believe they're part of the community. So they might do tree planting for example, with a school or whatever *because they're next door to that facility.*[84]

For instance, Austrian company OMV funded an annual Department of Conservation whale survey in Cook Strait for eight years until the oil market slump led to corporate cost-cutting. In the Taranaki region, Todd Energy has been a regular contributor to civic projects and community events. The company's donation toward a new aquatic center in New Plymouth was large enough to secure the company naming rights. During 2014 Todd Energy funded Todd Energy Rescue (a modern Coastguard rescue boat), the Todd Energy Raceway, the Taranaki Kart Club's raceway at Waitara, and the annual WOMAD (World of Music, Art, and Dance) festival. The company also made a $3 million contribution to the Len Lyle Arts Centre which opened in July 2015.

Not everyone in the community was impressed. Climate Justice Taranaki spokesperson Janice Liddle said the art center donation was "yet another case of the fossil fuel industry using their dirty money to buy social license/ acceptance within our communities. It's a pity that the sponsorship has come from such a destructive, unsustainable industry."[85] Another anti-oil activist maintained "this all means, of course, that the public and the New Plymouth District Council are firmly captured by Todd Energy's largesse. It's considered perfectly normal."[86]

Canadian company TAG Oil has undertaken a range of community initiatives "from formal partnerships, sponsorships and scholarships, to good-neighbor activities such as soundproofing and containing flare at drill sites at production stations."[87] Some of these donations have been targeted locally, like the
$25,000 for air conditioning, lighting, and a new kitchen for a school located
near one of its production sites. But the company also sponsors the Taranaki
rugby team in the ITM Cup, the Summer Scene holiday program in partnership
with Horizon Energy Services, and the New Plymouth Surf Life Saving Club

Scratch beneath the surface of this "social investment," however, and
there's occasionally a more calcualted agenda. As the editor of the *Taranaki
Daily News* suggested, perhaps oil companies are so liberal with their money
because it's necessary in order to retain their license to operate, unlike the
dairy industry.[88] For example, TAG Oil intensified its good-neighbor efforts
after the company was twice cited by the New Plymouth District Council in
2013 for non-compliance with noise ordinances and forced to close down its
operations.[89] The neighbors, with whom the company publically claimed it
enjoyed harmonious relations, had complained loudly enough to shut down
their operations at least temporarily.

A primary school in the small north Taranaki community of Tikorangi has
encountered similar problems. Tikorangi is in the middle of the unconventional oil and gas boom in the district, with fifty-five consented well sites,
drill rigs, and hundreds of daily truck movements. Freelance journalist Rachel Stewart was approached confidentially by a group of concerned school
parents in early 2013. They were worried about the effects of oil and gas
production near their school, and reports that Shell Todd Oil Services (STOS)
had approached the board of trustees offering $130,000 in "hush money" for
a school pool and solar water heating so drilling operations could proceed.[90]
The parents' concerns were borne out when the board eventually decided
to take advantage of the STOS funds while they were on offer, since it was
likely the drilling would proceed anyway.

Basil Chamberlain, Taranaki Regional Council chief executive, defended
the petroleum industry, claiming company donations and development levies
for new projects gave communities (and oil companies) something tangible to
point to. He had no doubt if the government decided to review the royalties issue, "companies would support returning revenue to the region if it *made business move more slickly and gave them more acceptance in the community.*"[91]

Infiltrating the Educational System

Globally, corporations in most business sectors seek to extend their influence
to the next generation by infiltrating the national educational system. The

petroleum industry is no exception, making extensive use of "educational campaigns," awards and "educational" materials (printed and online) to promote their industry to young people as well as to their parents and the general public.

New Zealand oil and gas companies have employed similar tactics through their communications or community relations offices, and indirectly through trusts, educational institutions, and other mechanisms. For example, Shell funds the Young Enterprise Scheme Taranaki, aimed at encouraging and supporting students to learn business and entrepreneurial skills. Canadian company NZEC supports the Western Institute of Technology's seventeen-week course offering a Certificate in Hydrocarbon Drilling. When the company secured an exploration permit on the East Coast in 2013, it announced six scholarships to East Coast students (mostly Māori) for tuition and living expenses at WIT. TAG Oil awards a $5,000 scholarship annually to a promising senior from Stratford High School who is interested in an oil- and gas-related field of study. The company's unconventional exploration and production operations are concentrated in the district.

GNS, a Crown research institute, works closely with the petroleum industry not only in providing scientific research services, but undertaking "educational" outreach and community engagement programs targeted particularly at school children, teachers, and iwi about geology and oil and gas development. For instance, at the 2016 New Zealand Petroleum Conference, petroleum geologist Dr. Kyle Bland described the agency's Te Kura Whenua (Earth School) initiative in conjunction with East Coast iwi Ngāti Kahungunu. Bland explained that the program was about "building relationships" and "partnering with iwi to achieve outcomes of long-term mutual benefit."[92] Media coverage of the program launch highlighted the fact that it not only gave participants a chance to learn about the geology of the area, but introduced them to oil and gas industry. The iwi's rohe (territory) covers some highly prospective geology on the East Coast.

Some of the industry's tactics have been decidedly dishonest, at least according to environmental critics. One of the more celebrated incidents occurred in early 2014 involving a so-called science roadshow titled "What Lives Down Under?" The show, which toured Taranaki schools, was the brainchild of a consortium composed of NZOG, TAG Oil, and Beach Energy. The consortium hired a purpose-built truck from the National Science-Technology Roadshow Trust and paid Wellington-based Somar Design Studios to design and develop an "educational" roadshow about geology, dinosaurs, and fossil fuels. The sponsoring oil companies' logos and company information featured prominently on the side of the vehicle and in the handout materials. According to TAG Oil, the roadshow was aimed at "harnessing kids' love of dinosaurs—and their unlimited imagination—to bring science to life outside of the classroom." The sponsors were "in the business of fossil hunting[!]" and it was a great way to "ignite kids' interest in geology."[93]

The New Zealand Greens party and Climate Justice Taranaki, whose representatives toured the display, accused the three oil companies of blatant propaganda. Climate Justice Taranaki's researcher Catherine Cheung said the materials "read more like pro-oil propaganda than scientific facts." The consortium, she claimed, used "the same old industry mantra: Everyone used oil and gas every day. We can't live in the modern world without it." It was an attempt to sell "what truly is a by-gone technology."[94] New Zealand Greens co-leader Dr Russell Norman agreed the show was clearly propaganda, and dodged the problems the industry created for society and the environment.[95] Education Minister Hekia Parata came to the defense of the oil companies involved, saying the government encouraged businesses to "share their knowledge and expertise [and funding] with schools in their local area."

"Partnering" with Tangata Whenua: Development or Disempowerment?

Indigenous populations have been subject to intense pressure from the petroleum industry to allow access to their land, often in remote and unexplored regions. The usual rhetoric is about building partnerships for economic and social benefit of the tribe, iwi or First Nation while preserving cultures and protecting local habitats. The tactics employed have become increasingly sophisticated. Communities often become divided between those who believe the companies' promises and those who see a threat to the tribe's autonomy and traditions.[96] Indeed, petroleum companies rely on such divisions and make every effort to stay on the side of leaders and factions who favor drilling, even if they are in a minority. For example, PhD researcher Keely Kidner[97] documented how the petroleum industry in Alberta, Canada, funded an Aboriginal Youth Task Force to run educational programs in order to get the majority of members of the Athabasca Chipewyan First Nation (ACFN) on-side with tar sands extraction and pipeline construction over their land. Much of ACFN territory overlies rich tar sands deposits that have been targeted by multinational oil and gas companies. ACFN is part of a network of first nations that has been fighting tar sands development because of its harmful ecological and cultural impacts.[98]

In New Zealand oil and gas companies hold out prospects of jobs or consultancy contracts as land access managers or "cultural facilitators" who assist in organizing "consultation" sessions to gain access to Māori land and public support for the company's activities. Many iwi leaders, particularly in the Taranaki region where the industry has been long established, put up with petroleum industry operations because it has become a fact of life and many of their people have obtained jobs and made careers as oil workers, truck drivers, and in service industries. STOS, for example, has contracted with a local hapū to provide drilling rig services and catering. They've hired members from another hapū to carry out cultural impact assessments and environmental monitoring. The resource consent process for new developments may be perfunctory, but since Māori are required to be consulted some leaders reason at least it gives their people get a seat at the table.

Others are more cautious, mindful of the divisions within their iwi and hapū as onshore unconventional wells continue to proliferate. Māori critics of the industry accuse some leaders of selling out, reiterating the traditional kaitiakitanga (guardianship) role of tangata whenua for the land.[99] They accuse companies like STOS and Statoil, with their training programs for marine observers and emergency accident responders, of neo-colonialism. Undaunted, the industry views such programs as useful PR illustrating partnership with Māori for mutual benefit. PEPANZ decided the partnership theme worth exploiting, and in 2015 contracted consultancy Spindletop Law to produce a research report on "Opportunities for Māori in the New Zealand Petroleum Sector." According to Spindletop, the purpose of the report was to suggest tools New Zealand's regulators could use to encourage greater Māori participation in the industry.[100] For PEPANZ, the report was a useful instrument for influencing policy as well as selling iwi on the idea of collaboration.

Te Rūnanga o Ngāti Ruanui, the commercial arm of a Taranaki iwi of the same name, entered into a partnership agreement with TAG Oil in December 2013. Signing the agreement TAG's CEO Garth Johnson said, "We want many of the same things: to be economically successful but never to the detriment of high environmental standards."[101] Ngāti Ruanui was promised employment opportunities and their environmental team would have archaeological oversight and monitoring rights for any earthworks TAG carried out in the region, in accord with the Rūnanga's Environmental Management Plan. TAG subsequently arranged for Ngāti Ruanui's chief executive to make presentations at two Advantage New Zealand conferences. Ruanui's environment manager also was invited to speak at the 2014 conference on "Best Practice Guidelines for Engagement with Māori." [102] The guidelines were later posted on government agency NZPaM's website.

Iwi "partners" have also proven helpful to oil and gas companies in addressing the concerns of other iwi and hapū over expanded exploration activities. After a new block offer was announced on the East Coast in 2013, Ngāti

Porou leader Sir Apirana Mahuika announced that the iwi was opposed to oil and gas exploration in their rohe. When the prime minister visited Gisborne in August 2013 to promote the idea of petroleum development, he claimed that unnamed Taranaki iwi were "big supporters of oil and gas exploration." They had been "over here talking with Ngāti Porou . . . and making the case that there are a lot of opportunities and a lot of jobs."[103] Apparently Ngāti Porou weren't persuaded by the discussions with their cousins. In December, Mahuika expressed anger over the Crown's lack of appropriate consultation before awarding of an exploration permit on the East Cape. It showed a blatant disregard for Ngāti Porou's mana whenua (authority over the land).[104] Government and company representatives had organized a few brief meetings with iwi representatives, some of whose leadership standing was in doubt, rather than arranging proper consultative hui so that all tribal members could be involved.

CONCLUSION

New Zealand oil and gas industry promotion has yet to reach the level of sophistication and resourcing of larger, more established petroleum-producing countries. Government's campaign to encourage and support expansion of the industry has struggled against determined and well-informed opposition, a weak international oil market, and rising concerns over climate change. Instead of a well-organized machine, the industry's promotion and defense proceeds through an informal network of personal and institutional relationships among government officials, industry executives, paid lobbyists, insider consultants, and a handful of supportive journalists and bloggers that has evolved under successive National governments. This diverse network shares a broad neoliberal ideology and hard-headed commitment to unfettered business growth, occasionally coordinates strategically, and includes some of the biggest corporations and most influential private-sector leaders in the country.

The New Zealand petroleum industry has extensive international connections and not surprisingly has copied certain strategies from Australia, North America, and the United Kingdom. For example, (a) the use of repetitious propaganda, (b) manufacturing "reasoned" debate to win over Middle New Zealand and respond to the emotive claims of environmentalists, and (c) co-opting communities and iwi through targeted funding, pseudo-consultation and unequal formal agreements. Opposition parties and environmental groups have been drawn into the debate in response to assertions that the most New Zealanders support oil and gas development because the industry creates jobs, is "environmentally responsible," and is leading the transition to a clean-energy future. To maintain the façade of 'reasoned' public discussion and dissemination of "factual" information, the industry has infiltrated the education system, funded academic centers and seminars, utilized supportive news sites and specialist consultants, exploited social media, and collaborated with business associations and think tanks.

That said, the country's distinctive history, relatively small population, legislative environment, and national culture traits have also influenced the strategies the petroleum industry has adopted. For example, the industry as a whole has placed considerable emphasis on community engagement, building personal networks, and establishing trust given the Kiwi propensity to eschew formality and look for practical solutions to problems through kahoi ki te kanohi (face-to-face) discussion. This style of approach is supported by the Crown Minerals Act which requires petroleum companies to consult and engage with local councils and Māori in a meaningful and culturally appropriate way.

Petroleum industry tactics to combat or neutralize organized environmental groups have undergone a gradual shift, from defense and confrontation toward respectful dialogue and seeking common ground. Three factors appear to have influenced this shift: (1) changes in strategy in overseas jurisdictions like North America, which have been conveyed to the New Zealand industry through professional contacts and by means of conferences like 2015 Advantage New Zealand which featured Tisha Schuller's presentation on "de-escalating the fracking wars"; (2) the New Zealand government's own shift in emphasis toward seeking cooperation with less extreme environmental and recreational groups in reforming legislation and implementing policy; and (3) the wind-back in onshore exploration activity and fracking due to the global oil market slump leading to fewer local protest actions while anti-fossil fuel campaigns by direct-action oriented environmental organizations continued.

The combined institutional effect of these industry and government tactics has been to sharpen the demarcation between those who are perceived as reasonable, responsible environmentalists "we can do business with" (through conference panels, media exchanges, symposia and various types of collaborative "partnerships"), and so-called extremist elements who refuse to engage in dialogue or compromise over environmental degradation and climate change, and are thereby marginalized by their own choices and actions (protests, civil disobedience, violence). The industry/government PR spin around this institutional realignment and marginalization of "green radicals" is crucial for keeping a large portion of Middle New Zealand on side and avoiding another upsurge in militant popular environmentalism as has happened in the past.

In the following chapter we will examine the tactics the National government used, in conjunction with the petroleum industry, local media, and business-oriented municipal councils to "sell" residents on the East Coast on the promised benefits of oil and gas development, and on the myth that they had a voice in whether such development took place.

NOTES

1. "Kiwi" is the colloquial term for a New Zealander. It refers to a small, flightless native bird.

2. Geoff Bertram, "Lessons from Think Big" (presentation to the New Zealand Association of Impact Assessment conference on Assessing the Impacts of Petroleum and Mineral Extraction in New Zealand, Wellington, December 10–11, 2012), emphasis added.

3. Cameron Madgwick, CEO of PEPANZ in an interview with the author on May 20, 2016.

4. See also Jane Mayer, *Dark Money: The Hidden History of the Billionaires behind the Rise of the Radical Right* (New York: Doubleday, 2016).

5. For example, ANZ Bank co-sponsors of the annual Advantage New Zealand Petroleum Summit and has increased its investment in gas as well as renewables. It's rationale for doing so is that the industry is, or at least purports to be, playing a lead role in the transition to a lower carbon economy. Personal communication from Stuart Watt, Contact Centres Team Leader, ANZ Bank, March 27, 2015.

6. "New Group Formed in Support of Drilling," *Otago Daily Times*, January 16, 2014, accessed September 18, 2014.

7. "New Group," *Otago Daily Times*.

8. Keely Kidner, "Beyond Greenwash: Environmental Discourses of Appropriation and Resistance" (PhD dissertation, School of Linguistics and Applied Languages, Victoria University of Wellington, 2015), 4.

9. See www.pepanz.com. The organization revised its website in late 2014, deleting the line "In all advocacy work we ensure that we are honest, reasonable and rational."

10. PEPANZ 2016 board members were Rob Jager, Chairman, Shell; Andrew Knight, Deputy Chairman, NZOG; Joanna Breare, Todd Energy; Max Murray, TAG Oil; Gabriel Selischi, OMV; Alan Seay, Anadarko; David Coull, Bell Gully; Nick Jackson, Elemental Group; and Maceon Cooper, Origin Energy.

11. "Oil Lobbyist—Between Bedrock and a Hard Place," *New Zealand Herald*, June 5, 2012, accessed August 2, 2014.

12. Geoff Bertram, "Lessons from Think Big."

13. "Oil Lobbyist," *New Zealand Herald*.

14. David Robinson, "New Zealand Petroleum Summit 2012 Launches into its First Day," Media release, PEPANZ, Wellington, September 19, 2012, www.pepanz .com.

15. Chris Bush, "Chairman's Address: Our Petroleum Future" (address to the New Zealand Petroleum Summit, Wellington, September 19, 2012), accessed August 4, 2013, www.pepanz.com.

16. Contrary to a 2014 Royal Society report that concluded transition to a low-carbon economy need not mean a significant decline in New Zealand's living standards.

17. "Distant Prospects," *Otago Daily Times*, Sunday Magazine, February 2, 2014, accessed July 25, 2015.

18. "Distant Prospects," *Otago Daily Times*.

19. Oil and Gas Potential Game Changer for New Zealand, *Scoop News*, July 26, 2016, accessed July 26, 2016.

20. "Oil Lobbyist," *New Zealand Herald*.

21. "Oil's not Well," *Otago Daily Times*, February 2, 2014, accessed July 10, 2015.

22. Chris Bush, "Chairman's Address: Our Petroleum Future."

23. Data provided in response to an Official Information Act request by Sarah Roberts of Taranaki Energy Watch to Brett Murray, General Manager High Hazards and Specialist Services, WorkSafe New Zealand, April 2015, File ref: 15/00203, http://www.epa.govt.nz/EEZ/EEZ000010/EEZ000010_Sarah_Roberts_Statement_of_Evidence_Climate_Justice_Taranaki_SubExWitness.pdf.

24. "Fracking safe, says Todd Energy," *New Zealand Herald*, November 8, 2012, accessed May 16, 2015. In their submission Todd sought to distinguish between the "shallow, water-hungry methods" used in the United States to source oil and gas from shale fields, and the fracking most commonly conducted in New Zealand. They neglected to point out that most of the limited fracking in New Zealand to date has been carried out in the tight sands geology of Taranaki. It was likely to be much different story with the shale-like formations on East Coast and elsewhere, as noted by the Parliamentary Commissioner for the Environment in her 2012 interim report on fracking.

25. Chris Bush, "Chairman's Address: Our Petroleum Future."

26. "PEPANZ Welcomes Rockefeller Divestment and Supports 'Transition'," *Radio New Zealand*, September 24, 2014, accessed October 25, 2014.

27. www.freemanmedia.co.nz.

28. E.g., David Coull, "Significant Period of Change."

29. David Coull, "Petroleum Business in New Zealand: the Continuing Challenge" (presentation to the Advantage New Zealand Petroleum Conference, Auckland, April 30, 2013).

30. David Coull, "Petroleum Business."

31. See www.kensingtonswan.com.

32. See archives on www.businessnz.org.nz.

33. See www.nzchambers.co.nz.

34. See Activate Tairāwhiti's "Economic Development Workplan 2015," http://www.activateTairāwhiti.co.nz/assets/Documents/Economic-Workplan-March-2015.pdf.

35. See http://nzinitiative.org.nz/

36. Jason Krupp, *Poverty of Wealth: Why Minerals Need to be Part of the Rural Economy* (New Zealand Initiative, Wellington, December 2014), https://secure.zeald.com/site/nzinitiative/files/Poverty%20of%20Wealth%20Final%20Web.pdf.

37. Jason Krupp, *From Red Tape to Green Gold* (New Zealand Initiative, Wellington, March 2015), https://secure.zeald.com/site/nzinitiative/files/Red%20Tape%20 final.pdf.

38. See Dunlop and McCright, "Organized Climate Change Denial," 4.

39. Nicky Hager and Bob Burton, "Secrets and Lies: How Shandwick PR Tried to Destroy the Rainforests of New Zealand," *PR Watch* 7, 1 (2000): 1–5.

40. McDouall Stewart, 2010, "Schedule 4 of the Crown Minerals Act is the Perfect Time for a Rational Debate in This Country," *Q&M Magazine* 7, 3 (2010), http:// contractormag.co.nz/contractor/time-for-a-rational-debate/.

41. Bernie Napp, "Fraction too Much Fiction about Fracking," Opinion, *New Zealand Herald*, August 8, 2011.

42. In their discourse analysis of 351 mainstream media stories on oil and gas development, Diprose, Thomas and Bond found that 10 percent of the articles included accusations that opponents were ill-informed (Ibid, Table 2, 165).

43. Chris Bush, "Chairman's Address: Our Petroleum Future," emphasis added.

44. PEPANZ, Job description: Resource Management and Community Affairs Manager, January–February 2014, www.pepanz.com, emphasis added.

45. "Security not Tight Enough at Oil Summit," *Taranaki Daily News*, October 1, 2014, accessed October 25, 2014.

46. "Security not Tight," *Taranaki Daily News.*

47. Interview with Cameron Madgwick, CEO of PEPANZ on May 20, 2016. The interviewee was advised the discussion was being recorded and consented to the information provided being available for possible publication.

48. "Global Campaign against Oil Operator Chevron Comes to NZ," *Voxy News*, July 22, 2015, accessed July 23, 2015.

49. David Robinson, "The Value of the Oil and Gas Industry to New Zealand" (presentation to the 2013 Advantage New Zealand conference, Auckland. April 28– May 1, 2013).

50. Simon Hartley, "Industry Sees Need for Community Engagement" *Otago Daily Times*, April 1, 2015; also "Oil and Gas Industry Locked into Programmes," *Otago Daily Times*, April 4, 2015.

51. Simon Hartley, "Oil and Gas Industry Locked."

52. Interview with Cameron Madgwick, CEO of PEPANZ on May 20, 2016.

53. Interview with a Taranaki environmental activist, December 5, 2015. The interviewee was advised the discussion was being recorded and consented to the information provided being available for possible publication on condition of anonymity.

54. "Massey Questions MPI Report," *Radio New Zealand* September 5, 2014. Also "Vet Warns of Oil Waste Farm Threat," *Radio New Zealand*, "October 7, 2014, accessed October 11, 2014.

55. Dr. Nick Smith, "Fracking Nonsense," PEPANZ newsletter, Issue 5, August, 2012, accessed November 12, 2015, http://www.pepanz.com/assets/Uploads/1863 -PEPANZ-Newsletter-August-.pdf.

56. http://www.gns.cri.nz/.

57. For example, "New Research Programme Aims to Lift Oil and Gas Exploration," *Scoop News*, September 3, 2015, accessed September 5, 2015.

58. "GNS Scientists Speak Out on Oil and Gas Exploration," *Māori Television*, November 13, 2015. See also "Quake Risk from Drilling Tiny," *Gisborne Herald*, November 14, 2015, accessed November 14, 2015.

59. Interview with Chris Hollis of GNS on *Te Kaea*, November 13, 2015, accessed November 14, 2015.

60. See www.eetnz.org.nz.

61. The Royal Society of New Zealand, *Transition to a Low-Carbon Future for New Zealand*, Wellington, April 2016, http://www.royalsociety.org.nz/media/2016/04/Report-Transition-to-Low-Carbon-Economy-for-NZ-1.pdf.

62. Frank Duffield, "Natural Gas can Smooth Transition: Turning too Quickly to Renewable Energies Creates Precarious Supply," *New Zealand Herald*, May 13, 2016, accessed May 14, 2016.

63. Interview with Cameron Madgwick, CEO of PEPANZ on May 20, 2016, emphasis added.

64. "Oil, Gas Sector Told to Improve Communication," *Taranaki Daily News*, October 10, 2014, accessed August 3, 2015.

65. Communication with a leading New Zealand anti-oil activist who administers several Facebook pages, May 8, 2014.

66. Herman and Chomsky, *Manufacturing Consent*, 26 ff.

67. Hager, *Dirty Politics*, 16ff.

68. See presentations at the 2011 Houston Unconventional Oil and Gas Industry PR conference discussed earlier.

69. "Concerns Mount over Anadarko Drilling," *TV3 News*, October 23, 2013, accessed August 15, 2014.

70. Terrence Loomis, "Concerns over Oil and Gas Waste," op ed piece, *Gisborne Herald*, September 16, 2013, accessed September 16, 2013.

71. Nicky Hager, *Dirty Politics.*

72. "Rodney Hide on Dunedin's Luddite Council," *Whale Oil* (blog), May 24, 2014, accessed August 5, 2015, http://www.whaleoil.co.nz/2014/05/rodney-hide-dunedins-luddite-council/.

73. "Something More for the Greenies and Labour Party to Cry About," *Whale Oil* (blog), November 8, 2012, accessed July 14, 2015, http://www.whaleoil.co.nz/2012/11/something-more-for-the-greenies-and-labour-party-to-cry-about/.

74. "Limited Coverage of Proposed New Marine Protection Framework," *Scoop News,* January 12, 2016, accessed January 13, 2016.

75. C.D. Baker and J.B. Napp, "Minerals Briefing Paper 2014" (Straterra Inc, Wellington, April 2014), www.straterra.co.nz. The subtitle of the paper was "Policies for Increasing New Zealand's Attractiveness for Investment in *Responsible Minerals Exploration and Mining*," emphasis added. Also see "Straterra Minerals Briefing Paper Launched at Parliament," *Scoop News*, April 9, 2014.

76. See www.freemanmedia.co.nz.

77. "Minerals Key to Waikato-Tainui Prosperity – MP," *Stuff News*, October 23, 2013, accessed September 30, 2014.

78. "Shane Jones Takes $$ from Energy Sector," *Newstalk ZB*, April 22, 2014, accessed September 30, 2014.

79. Interview with Cameron Madgwick, CEO of PEPANZ on May 20, 2016.

80. Simon Bridges, "Speech to the Advantage NZ 2013 conference."

81. Interview with Josh Adams, NZPaM National Manager Petroleum on December 2, 2015.

82. "TRC Boss Denies Claims of Conflict," *Taranaki Daily News*, March 12, 2012, accessed December 2, 2014.

83. See www.nzog.com.

84. Interview with Cameron Madgwick, CEO of PEPANZ on May 20, 2016, emphasis added.

85. Climate Justice Taranaki, Media release, "Len Lye Arts Centre and the Fossil Fuel Industry," New Plymouth, July 24, 2015, https://climatejusticetaranaki.wordpress.com/2015/07/24/press-statement-len-lye-arts-centre-and-the-fossil-fuel-industry/.

86. Rachel Stewart, "When the Oil Folks Come to Town," *Manawatu Standard*, April 2, 2013, accessed August 10, 2015.

87. See www.tagoil.com/new-zealand-community/.

88. "Editorial, "Fuel for Thought," *Taranaki Daily News*, September 6, 2014, accessed August 10, 2015.

89. "Noise Silences Sidewinder Drilling," *Taranaki Daily News*, June 4, 2013, accessed January 11, 2015.

90. Stewart, "When the Oil Folks Come to Town."

91. "Local Slice of Oil, Gas Royalties Advocated," *Taranaki Daily News*, September 1, 2014, accessed October 10, 2014, emphasis added.

92. Kyle Bland, "Partnering with Iwi to Achieve Outcomes of Long-term Mutual Benefit" (presentation to the New Zealand Petroleum Conference 2016, Auckland, March 20–22), http://www.petroleumconference.nz/presentations-online/.

93. TAG Oil, February 20, 2014, www.tagoil.com

94. "Pro-oil Propaganda not Science," *Voxy News*, February 16, 2014, accessed September 20, 2014.

95. "Greens Claim Roadshow Propaganda," *Radio New Zealand,* February 17, 2014, accessed September 20, 2014.

96. See Patricia Widener, "Benefits and Burdens of Transnational Campaigns: A Comparison of Four Oil Struggles in Ecuador," *Mobilization: An International Quarterly* 12, 1 (2007): 21–36.

97. Keely Kidner, "Beyond Greenwash," 190ff.

98. For more information, see ACFN's website http://www.albertatarsandsnetwork.ca/athabasca_chipewyan_first_nation.

99. Author's interview with a Taranaki Māori environmental activist, February 9, 2016. The interviewee was advised the discussion was being recorded and consented to the information provided being available for possible publication on condition of anonymity.

100. Spindletop Law, "Report on the Opportunities for Māori Participation in the New Zealand Petroleum Sector" (Spindletop Law, Wellington, October 2015), http://www.spindletop.co.nz/spindletop-report-on-the-opportunities-for-Māori-in-the-new-zealand-petroleum-industry/.

101. "Community Counts," TAG Oil, accessed August 21, 2015, http://www.tagoil.com/about/new-zealand-community/.

102. See Sarah-Lee Rangi, "Best Practice Guidelines for Engagement with Māori," Te Rūnanga o Ngāti Ruanui, Taranaki, 2013, https://issuu.com/sarah-leerangi/docs/best_practice_guidelines_for_engage/1.

103. "PM: Mining, Oil Huge Potential for Region," *Gisborne Herald*, August 29, 2014, accessed October 15, 2014.

104. "Exploration Permits 'Show Disregard' for Ngāti Porou," *Gisborne Herald*, December 7, 2013, accessed September 17, 2015.

Chapter Five

Selling the East Coast

New Zealand's East Coast[1] has been one of the first regions to be targeted for expansion of onshore and offshore oil and gas development by John Key's government. A "game-changing" discovery outside of Taranaki would signal to major international oil and gas corporations and their investors that the country has serious prospectivity. In addition to an international marketing campaign, free geological data, subsidies, and promotional conferences, the government has spent considerable time and resources selling East Coast councils, iwi, businesses, and communities on the idea of oil and gas development.

This may seem odd, since under current government policy settings East Coasters have little influence over such developments other than at the ballot box. But this is one reason *why* selling the region on petroleum development has been high on the government's political and Business Growth Agenda. As a struggling rural-based economy with a high Māori population, the East Coast has been subject to various government interventions over the years. It was an ideal test case for how the government could engage key stakeholders (including cautious farmers and iwi) and convince reluctant local residents to buy into the oil and gas industry while rolling out a new regional development program. National placed considerable emphasis on appearing to consult, inform and encourage "reasoned" debate because widespread acceptance of such a high-impact, environmentally risky industry would help marginalize environmentalists and anti-fracking groups and boost support for government's Business Growth Agenda. The choice needed to be seen as involving East Coast residents, not just driven by the government.

Most district councils on the East Coast were dominated by business and agricultural interests. When push came to shove they backed any kind of development that would generate growth, jobs, and more rates (taxes). Given all the government hype about the potential of oil and gas development, what they *weren't* prepared to brook was such a "promising" industry being blocked by what many mayors, councilors, and business people perceived (and portrayed in public meetings and the media) as a minority of tree-hugging leftist extremists and Māori radicals. In some cases, as we will see, council actions went beyond industry cheerleading to actively preventing industry critics, environmental activists, and the wider public from having a say about oil and gas development and resource consents.

In this chapter we will examine the manoeuvers employed by ministers and state officials, in conjunction with the petroleum industry and municipal authorities, to sell East Coasters on the industry and counter environmental opposition and community anti-fracking groups.

GOVERNMENT'S PROMOTIONAL MANEUVERS

The history, demographic trends, and underdevelopment of the East Coast provided government politicians and oil industry representatives with ammunition to build a case for oil and gas expansion. But a variety of economic and social factors, as well as the emergence of a "flaxroots" anti-petroleum/anti-fracking movement, help explain why selling the East Coast has so far met with a mixed response. The population of the East Coast of New Zealand has been fairly static at around 180,000 since the turn of the twenty-first century. The region has a greater proportion of youth compared with the rest of the country and lower numbers of people of prime working and childbearing age. These people often have to go elsewhere for steady employment particularly if they're seeking a career in business, professional services or government. Ethnically the East Coast is 64 percent European (67 percent nationally), 27 percent Māori (13 percent nationally) and 9 percent Pacific, Asian, and other (20 percent nationally including 11 percent Asian). The region is heavily influenced by Māori culture. GDP was $7.8 billion in 2013, contributing approximately 4 percent to national GDP. The main productive sectors are forestry, horticulture (including wine), livestock and crop farming, and fishing.[2]

Disseminating "Factual" Information

The government wasted no time once it decided to target the East Coast for regional development, including establishment of an oil and gas industry. In 2012, with a new round of exploration block offers underway, MBIE's Economic Development Group began work on an occasional paper on the economic potential of the oil and gas industry for the country.[3] In parallel with that paper, they commissioned two East Coast petroleum development studies by independent consultants.[4] By the beginning of 2013, a ministry team was at work combining the studies into a report on the potential for oil and gas development on the East Coast. All the papers received extensive input from the petroleum industry.

The problem was that it was MBIE's report. When it was released, it could give the appearance that the government was trying to tell East Coasters what to do. So at the invitation of Minister Joyce, a steering group of mayors and regional council chairs was convened in early 2013 and asked to lend their support to the "joint study." It was a request they could hardly refuse; the study was already underway. Establishing the steering group gave the government's report legitimacy. It also gave the local councils and national government representatives a vehicle to claim they were objectively investigating the "potential benefits, impacts and risks of petroleum (oil and gas) development" in order to "support *informed dialogue* between councils, communities and iwi."[5] In reality, it was about generating public discussion so the people of the East Coast believed they actually had a say in the matter. Government policy regarding oil and gas development was already set in place.

Three months later, Economic Development Minister Steven Joyce launched the first public salvo in the government's sales campaign with the release of the East Coast Oil and Gas Development Study.[6] It certainly achieved its aim of generating debate, judging by the rash of newspaper editorials, letters to the editor, op-ed pieces, and public forums that followed.

The MBIE study deserves closer examination because it provides a glimpse into how government with assistance from the petroleum industry went about estimating of how much resource there might be on the East Coast, whether there could be a "boom," what the potential benefits (and risks) might be, and packaging all this in a way that would sell oil and gas exploration to the people of the region. The study set out five development scenarios, two of which were perfunctorily rejected as uneconomic. That left three which even the MBIE authors acknowledged needed to be treated with caution because they were so speculative:

> Scenario 3: Small-scale production, with oil produced in three areas of the East Coast over a period of twenty-one years.
> Scenario 4: Large-scale production over a period of forty-one years that "would have a significant impact on both the regional and the national economy."
> Scenario 5: Large-scale, high-volume production over a period of sixty-four years; the authors considered it to be at the "very upper end of what may be plausible, but with potential transformative implications."[7]

The fifth scenario could result in a "significant oil boom" that the study team claimed could be "game changing for New Zealand as a whole."[8] The main benefits claimed for the scenarios are set out in table 5.1.

Table 5.1.   Benefits of East Coast Oil and Gas Development.

Benefits	Scenario 3	Scenario 4	Scenario 5
GDP growth	$360 million	$5.3 billion	$18 billion
Crown revenue	$88 million	$2 billion	$7.7 billion
Direct employment (net)	158	587	899
Industry value added:			
Mining services	2.5%	18%	48%
Heavy construction	.22%	1.7%	4.7%
Supermarkets	.10%	1.3%	4.5%

Source: "East Coast Oil and Gas Development Study" published by the Ministry of Business, Innovation and Employment, 6–8 (Wellington: New Zealand Government, 2013). Copyright owned by the Ministry of Business, Innovation and Employment on behalf of the Crown.
   Unless indicated otherwise for specific items or collections of content (either below or within specific items or collections), this copyright material is licensed for re-use under a Creative Commons Attribution 4.0 International License.

The report listed the possible negative impacts of what was likely to involve unconventional oil and gas development as infrastructure, water use, methane, and other air emissions. Other problems (documented in the international literature discussed previously) such as waste treatment, risks of groundwater contamination, increased seismicity, and infrastructure impacts were ignored. The report made no attempt to quantify the negative impacts for each scenario, as it had done with the claimed benefits. Instead these were described in general terms and illustrated with a few examples from New Zealand experience, which were obviously selected to downplay their significance.[9]

Whether the East Coast might experience a "game-changing" boom depended on the volume of commercial reserves under the region and on the state of world markets. Data on proven reserves in the basin was sketchy so the MBIE team resorted to scenarios. The reliability of these scenarios depends on which analogue (a comparison producing field) was chosen for comparison, how closely it approximated the geology of the East Coast and how rigorously the study team applied evidence from the comparative field. The MBIE study cited TAG Oil's use of the unconventional Bakken oil and gas field in North Dakota as a rationale for their scenarios, since TAG had permits to explore on the East Coast. Data from unconventional US fields like the Bakken, Eagle Ford, and Barnett suggest the MBIE study significantly underestimated (or ignored) how fast unconventional wells deplete and how quickly any boom can go bust. Production from the average Bakken well declines 69 percent in the first year. By the end of five years, the average unconventional well's productive life is entering a steep decline, so companies have to drill more and more wells just to maintain production.[10] Industry data indicates the average Bakken well produces around 300,000 barrels (bbls) of oil equivalent over its lifetime, only a third of the recovery rate estimated by TAG Oil for its East Coast operations using Bakken as their model (and used by MBIE in their scenarios). At that rate, it was unlikely even MBIE's scenario 5 would produce the "game-changing" boom the study team anticipated. It didn't auger well for their claims about the economic benefits to the region.

The economic contribution of oil and gas development to the East Coast also depended on how well the MBIE study team understood and interpreted various terms the industry used in estimating and reporting potential reserves. Below is a diagram that might help in clarifying oil industry terms. It is from the Society of Petroleum Engineers "Guide for Non-Technical Users" for understanding the international Petroleum Resources Management System that oil companies and resource assessors use. The further up the diagram one goes, the more the estimated volumes in a given field move from educated guesses of undiscovered resources, to discovered resources, to reserves that can be commercially developed. However, the volumes also get smaller the further up the chart one goes. *Proven commercial reserves* are the smallest amount of all. It means somebody actually drilled down and sampled how much resource there was that could be profitably extracted.

TAG Oil, whom the study team relied on for much of its information, claims on its website that its East Coast holdings had "world-class upside potential" with "billions of bbls of *potentially recoverable oil*." But Sproule International, who carried out an independent resource assessment for TAG,

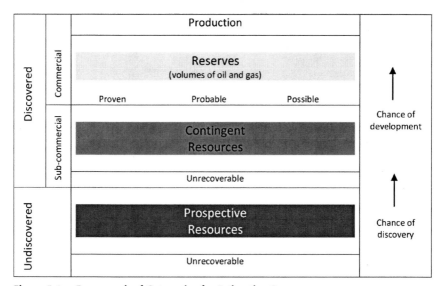

Figure 5.1. Framework of Categories for Estimating Reserves.
Source: Table "Categorize based primarily on techincal uncertainty of sales quantities associated with a project," in *SPE Petroleum Resources Management System Guide for Non-Technical Users, 2*. © Society of Petroleum Engineers International, 2007.

stated: "It is our opinion that the basin is too immature to justify estimation of recoverable hydrocarbons, thus we have limited our assessment to the estimation of *undiscovered in-place resources*."[11] The study team appears to have accepted the more optimistic projections from TAG and other companies in its scenarios.

In conclusion it seems highly unlikely the East Coast will experience the kind of boom that has occurred in the Bakken, Barnett, or Eagle Ford shales in the United States. The MBIE study was considerably wide of the mark about production and projected benefits. Since then, the global oil glut has delayed the exploration plans of junior and middle-range companies, most of whom have at least temporarily withdrawn from the East Coast. Even after the oil market has recovered, the East Coast is likely to see fairly limited exploration and at most a handful of producing wells for some time. It appears critics of the report were right to be skeptical about oil and gas development on the East Coast.

While the 2013 oil and gas study was underway, MBIE initiated another study called East Coast Regional Economic Potential. This was the first in a series of regional studies under the direction of Economic Development Minister Steven Joyce timed to coincide with the lead-up to the 2014 election. They were part of government's campaign to demonstrate its commitment to the regions and sell the East Coast on oil and gas development. A steering group composed of the same local government representatives as the 2013 East Coast Oil and Gas Development Study, with the addition of the Ministry of Transport and key iwi leaders, was convened in May 2013. The expanded membership once again helped legitimate the study as a "collaborative partnership" rather than an economic development directive from central government. [12]

Stage 1 of the study, a literature review, referred to oil and gas development as a "more speculative opportunity," no doubt in acknowledgment of criticisms the earlier 2013 study had received. The authors cautioned the scenarios in the 2013 report had yet to be tested by exploration and drilling, and that

> public and iwi concerns are still to be addressed. . . . The [2013] oil and gas study did not aim to provide any recommendations or forecasts. It simply explained the *significant choices that may need to be made and the existing mechanisms for the people of the East Coast to provide input into any decisions.*[13]

The language implied that oil and gas development was not a fait accompli and there were processes by which East Coast residents could help make "significant choices." There weren't, other than national elections, but few challenged the report's claims. Newspaper editors, business leaders and local mayors picked up the theme that people should inform themselves and get involved in the "public debate."

Stage 2 of the report[14] involved economic development forecasting with specific attention to transport and employment issues, although half the report was devoted to oil and gas development. Like stage 1, it built on MBIE's 2013 study to "forecast the potential impact that oil and gas developments may have on the region."[15] Ministers and local media made much of stage 2 to try to generate public discussion and engender support for the petroleum industry, so it is worth examining more closely.

After suggesting three East Coast economic growth options (low, baseline, and high), the study team reviewed oil and gas scenarios 3 through 5 from MBIE's 2013 study. They then took the extraordinary step of arbitrarily deciding "not to model scenario 5 as it represents an extreme case, i.e., the upper end of what is plausible."[16] This is bizarre reasoning, since the scenarios were originally compiled by experienced officials with input from the petroleum industry, GNS Science and two respected consultancies. Scenario 5 was precisely the kind of "boom" the industry and ministers had been holding up as a realistic prospect for the East Coast.[17] The only sensible explanation for rejecting scenario 5 seems to be that the study team already knew the figures behind the 2013 report scenarios were wrong, and the potential negative impacts (water demand, waste disposal, truck traffic, environmental damage) of scenario 5 with its thousands of fracked wells might frighten the public and provide further ammunition for environmentalists.

That left scenarios 3 and 4, which contained fewer wells, more modest output, and fewer negative impacts. It also reduced the claimed benefits. Fewer benefits were a problem for government ministers and industry representatives who were hoping to generate support for expansion of the industry. So the team resorted to manipulating some of the assumptions from the earlier 2013 study and forecasting almost double the number of jobs.[18] Even so, the team acknowledged "many of the required occupations would need to be filled by people from outside the study area."[19] They also noted that export-oriented industries could experience a reduction in value-added due to appreciation of the New Zealand dollar, tourism may suffer, and road transport would be negatively affected. Readers could rightly question whether the claimed "benefits" were believable and worth the costs.

A month after the release of the East Coast Economic Potential Study, NZ-PaM officials visited Gisborne for a confidential meeting with Gisborne District councilors to discuss the idea of holding a multi-agency public information hui on government's oil and gas regulatory regime. Like other NZPaM roadshows, its underlying aim was to reassure the public that an expanded oil and gas industry was safe and economically beneficial. The Council meeting excluded the public so that protocols and procedures for managing the hui could be discussed,[20] no doubt taking a lesson from NZPaM's prior experiences with critical audiences in Kaikoura and Dunedin.

The hui was held the evening of July 15, 2014, in conjunction with the Gisborne District Council's long-term planning exercise, and attracted over a hundred people. A group of approximately twenty protesters, mostly Māori women, assembled outside the venue displaying banners and chanting slogans against oil and gas exploration. NZPaM's Iwi Relations Manager Pieri Munro, an ex-policeman, hovered near the entrance keeping an eye on the protestors and ensuring that people arriving weren't prevented from entering. The program was tightly structured and moderated by an independent local Māori chairperson. NZPaM's GM James Stevenson-Wallace began with a much truncated slide-show explaining the stages of exploration and development and the country's regulatory regime. Wellington officials then spoke briefly about their agency's role in administering government regulations, after which the floor was open for questions. At least six individuals, including a conservative district councilor known to back the industry, asked virtually the same question about the benefits of oil and gas at different points during the discussion. Most of the questions and comments were about the impacts of oil and gas development, not about government regulations. Officials seemed to anticipate this from previous experience, and spent most of their time reassuring people about the robustness of the government's regulatory regime and safety of the petroleum industry. For example, NZPaM's chief geologist responded to concerns about fracking chemicals by praising the transparency of the industry and mentioning the establishment in the United States of the website FracFocus, which supposedly enabled companies to report well locations and the chemicals they used in fracking. He surmised a similar voluntary database might be set up in New Zealand. I took the opportunity to point out that FracFocus had been criticized by environmental groups, who found that most companies didn't report all their wells and declined to disclose all their fracking chemicals with the excuse they were commercially sensitive. Toward the end of the meeting, an official responded to several questions about the effects of fracking by reiterating New Zealand's rigorous and "world class" regulatory regime. I rose and suggested he may want to read a 2014 open letter to New York governor Andrew Cuomo from over one hundred health professionals and scientists citing peer reviewed studies of the harmful effects associated with fracking regardless of how rigorous the regulatory regime. When he responded skeptically (not doubt for the benefit of the audience) that he'd like to SEE this evidence, I went to the front and handed him a copy.

After the hui, Stevenson-Wallace dismissed criticism that the gathering was a "sales pitch" to promote oil and gas exploration.[21] Local New Zealand

Greens candidate Gavin Maclean disagreed in an op-ed piece in the local paper the following weekend,[22] accusing officials of not listening to local concerns. The government, he asserted, was ignoring the risks of offshore and unconventional oil and gas development and reneging on its promises to combat climate change. It was only three months before the national election, so it was hardly surprising (except perhaps for the close timing) that Maclean's piece was accompanied by a rebuttal letter from Energy Minister Simon Bridges.[23] Bridges began with standard government cant about the NZPaM meeting and Maclean's opinion piece being an opportunity for "well-informed debate." The remainder of his letter was a series of his stock platitudes (see chapter 3) about the petroleum industry's safe and environmentally responsible operations, government's "world class" regulatory regime, and the jobs and growth the industry would bring to the East Coast.

Unbeknown to Maclean, the editor of the *Gisborne Herald* had adopted a practice of delaying critical letters to the editor or opinion pieces until the relevant minister, official, or industry representative had had a chance to read them and prepare a response. The items were then published simultaneously as a kind of phony debate, first the critical piece and then "factual" authoritative rebuttal. It was a tactic straight out of the 2011 Houston PR conference about "harnessing the media" and was patently unethical, because it gave those in privileged government or corporate positions an unfair advantage over ordinary citizens. The government facilitated such arrangements through what was termed "all of government media monitoring," a practice defended by the government as essential for officials and ministers to do their jobs.[24] In October 2015 it was revealed that super-ministry MBIE spent $800,000 over two-and-a-half years monitoring print and broadcast media throughout the country.[25]

Ministerial Jawboning[26] and Cheerleading

In August 2012, Economic Development Minister Steven Joyce asked representatives of East Coast councils to come to Wellington for discussions on regional development. Energy and Resources Minister Phil Heatley was in attendance and most of the meeting focused on oil and gas development. According to a report by Gisborne District Councilor Andy Cranston, who represented the mayor and council at the meeting, Joyce strongly "lobbied" East Coast Councils to get behind the National government's oil and gas policy.[27] Joyce apparently felt councils were not doing enough to show their public support for establishing the industry in the region. The Ministry had already begun work on the East Coast Oil and Gas Development study in conjunction with an economic consultancy, and both ministers proposed forming a steering group to demonstrate local involvement and support for the process. Some of local government representatives expressed concern that the government was putting all the responsibility on them, and not doing enough to sell the benefits of oil and gas to East Coasters. The Ministers agreed to visit the East Coast more often to "campaign" for oil and gas development as well as arranging visits by other politicians and possibly the PM. In turn, the local government representatives committed to participating in the "independent study" of the socio-economic benefits of oil and gas, and *publically backing the report's conclusions*—even before the report had been completed.[28]

Joyce's promised campaign got under way with the release of the MBIE East Coast Oil and Gas Development study at the beginning of March 2013, shortly before the announcement of the latest exploration block offer that included parts of the East Coast. Ministers Joyce and Bridges (recently appointed to replace Heatley in the Energy and Natural Resource portfolio) visited Napier/Hastings and Gisborne to launch the study and promote the oil and gas industry. Both ministers emphasized the importance of the public being involved in discussions and in any decisions about the future development of the industry. Given the block offer process already underway enabling E&P companies to bid for exploration permits, the ministers' comments appeared to be little more than an attempt to hoodwink the public.

A month later, Bridges was back in Napier for a closed meeting with the region's mayors, chairs, councilors and council staff (many of whom were on the steering group for MBIE's just-released study) as well as members of the Hawke's Bay Chamber of Commerce. An exploration permit covering much of the productive sheep and beef farming and wine-growing area around Napier/Hastings had been granted in 2012. Rural landowners were understandably nervous about the possible downsides of oil and gas development. Lobby group Don't Frack the Bay organized a rally outside the meeting venue, complaining that the general public had been excluded after Bridges and Joyce had earlier promised open discussions with the whole community.

In May 2013, as promised earlier by Minister Joyce, Prime Minister Key visited Napier to speak at a closed meeting involving most of the same individuals and organizations Bridges had met with a month earlier. Key claimed the oil and gas industry would boost Hawke's Bay's economy like it had in Taranaki. "There have been no environmental problems in Taranaki and Hawke's Bay should be no different," he said. Key's claim was factually incorrect (there had been over a hundred officially reported incidents) though no one in this carefully chosen audience was likely to challenge him. Following the PM's visit, Don't Frack the Bay organized several protests over the next few months as well as "information" meetings for rural landowners to discuss the dangers of unconventional oil and gas development.

In early November following national protests organized by Greenpeace over deep-sea drilling, Minister Bridges decided to post an opinion piece to regional papers on the safety of oil and gas operations. He assured people who were concerned about onshore and offshore drilling that the government "shares your desire to protect and respect your local environment."[29] He emphasized that he supported development of oil and gas "in a sensible, safe and environmentally responsible way. *This is certainly not about development at any cost.*" Bridges welcomed the Parliamentary Commissioner for the Environment's (PCE) interim report on fracking and pointed to the government's tightening of safety regulations. He ended his piece with one of his (and the industry's) favorite mantras that expanded petroleum development would create jobs and provide revenue for hospitals, schools, and roads. As one of the 2011 Houston petroleum PR conference speakers enjoined his listeners, "Don't forget—It's jobs, jobs, jobs!"

A week later a letter to the editor appeared in the *Gisborne Herald* from Pauline Doyle of Napier,[30] taking Bridges to task over his comments about the oil and gas industry's safety and environmental record. Citing the PCE's interim fracking report, she noted that until recently the Taranaki Regional council didn't require drilling consents and did not monitor environment effects. It was left to the industry to do their own monitoring and reporting. If there had been any problems with fracking, Doyle claimed, the industry simply didn't need to report them; nobody would know. True to form, the editor of the *Gisborne Herald* withheld publication of Ms Doyle's letter for several days until the minister (or more likely his press secretary) had a chance to read it, consult the Taranaki Regional Council, and prepare a response. Bridges's rejoinder was published on the same day as Ms. Doyle's letter, defending his earlier statements and praising TRC's regulatory procedures.

At the beginning of November 2013, Minister Joyce made an unannounced trip to Gisborne for confidential discussions with the Chamber of Commerce, the Eastland Energy Trust and the Gisborne District Council about government's plans to push regional development.[31] A month later, Joyce was back addressing a Chamber of Commerce economic development forum for members only and invited media. Once again he engaged in fabricating "reasoned" public debate, urging "Gisborne people to explore the thought" and "have a look" at how oil and gas exploration could transform the region as it had done in Taranaki.[32] Unfortunately, Joyce neglected to mention the boom and bust nature of the industry. Eighteen months later, oil and gas companies were laying off workers and Taranaki was reportedly the weakest-performing region in the country.[33] Labour and National politicians alike, *including* Minister Joyce, were calling on Taranaki to diversify its economy.

During his December 2013 visit, Joyce also mentioned that he had met with Ngāti Porou leaders and "discussed the potential for oil and gas development with them."[34] The minister was clearly working hard to preserve the illusion that the public and iwi had a say in whether oil and gas development occurred. Unfortunately, that effort received a setback when, the day after Joyce left Gisborne, Te Rūnanganui o Ngāti Porou chairman Sir Apirana Mahuika accused the government of "blatant disregard" for the mana whenua rights of iwi over the way the Crown was handling exploration permit consultations.[35] Mahuika maintained that government officials had only met with a handful of iwi representatives. This wasn't the way things were done among Māori: "Each iwi has a right to be properly consulted. Every individual Ngāti Porou land owner and or shareholder in land has to be heard rather than have others make a determination on their behalf."

Joyce returned to Gisborne four months later in April 2014 with a delegation including Ministers Gerry Brownlee and Ann Tully (also the local National Party MP) to release the two-stage East Coast Economic Potential Study at another invitation-only gathering. With the national election only five months away, ministers clearly wanted to avoid protesters disrupting the proceedings.

Prime Minister Key visited the East Coast again at the end of August 2014 during the election campaign, emphasizing the region's need for stronger economic growth. Like Taranaki, he believed the East Coast's comparative advantage could include oil and gas.[36] It also appeared the government had been working behind the scenes to smooth over Ngāti Porou's concerns regarding the Crown's consultation processes. Key said he was aware of concerns among local iwi. Taranaki iwi had been "over here talking to Ngāti Porou and really making the case that there are a lot of opportunities and a lot of jobs. . . . The reality is Taranaki iwi are one of the biggest supporters" of oil and gas development.

In October 2015 after National had been re-elected to a third term, Joyce and his department were back on the marketing offensive aided by the local Gisborne paper. A just-released MBIE Regional Economic Activity Report showed that the region ranked "dead last" in economic performance among the regions.[37] Like previous MBIE studies, the report pointed to "untapped oil and gas reserves" as a prime opportunity to boost the East Coast economy. Joyce announced the government was working with economic development agency Activate Tairāwhiti to create a growth plan for the region:

It's important that the region gets together as a whole, including the Māori economy, and says *this is our plan* for the Tairāwhiti region—that this is the greatest chance for moving the dial for Gisborne and the whole Tairāwhiti region.

Joyce may well have enjoined the public to get behind the plan, but they weren't about to have a say in its development nor was the local Council. Whereas he had virtually pressganged local mayors into supporting the government's regional economic studies, now Joyce directed MBIE to only involve the business sector and commercially minded iwi. The local council and the general public were to be excluded until the report was completed. Critics claimed the process over-rode the Gisborne District Council's long-term community plan processes (which involved the public) and addressed transport, land use and social issues associated with development.[38]

BIG OIL'S EAST COAST MARKETING STRATEGIES

By the end of 2012, MBIE and PEPANZ had already laid the groundwork for companies with East Coast exploration permits to launch their own promotion and "consultation" activities. Most of the players were small to medium-sized Canadian companies who either had in-country executives to do their PR or contracted New Zealand-based communications people as needed. Over succeeding months, these companies and PEPANZ cooperated in keeping track of local political developments, monitoring protests, scheduling visits to the East Coast, and developing promotional materials and media responses to environmental activists and concerned citizens.

Promotion, PR Spin, and Use of "Experts"

The most active corporate player and promoter of oil and gas exploration on the East Coast has been Canadian company TAG Oil. Their operations are largely centered in the Taranaki Basin, but in recent years they have undertaken preliminary exploration activities on the East Coast. As an overseas-based company, TAG had faced difficult public relations challenges having to address diverse audiences: shareholders, the New Zealand government, tangata whenua, the general public, and environmental protesters. As CEO Drew Cadenhead told an interviewer in 2014, the process hadn't been easy because the industry had come under increased public scrutiny. Maintaining the myth that oil companies in New Zealand needed to earn "permission and consent to drill wells," he said TAG had put in "many, many more times the effort" in community relations than a few years ago.[39]

TAG obtained East Coast exploration permits in 2006, but didn't lodge their first resource consent applications until the beginning of 2012. In March 2013 TAG had received resource consent from Tararua Regional Council to drill two exploratory wells near Dannevirke in the south of the East Coast.

CEO Garth Johnson assured the public no fracking would take place (at least initially), and the company would "carry out all work on the East Coast in the *same careful, methodical and safe way* that we have worked in Taranaki for over a decade."[40] TAG began work almost immediately on the well site and was drilling within a month. Local landowners were not entirely convinced by Johnson's PR spin. Several staged a protest at the site, complaining of truck traffic, threats to land values and the company's lack of consultation.[41] The following day, another group of protesters marched on the site warning of pollution risks to groundwater and district waterways.[42] CEO Johnson responded by encouraging people to read a recent Taranaki Regional Council monitoring report on two TAG well sites that found no evidence of groundwater contamination in relation to these "world class" operations.[43] He acknowledged the concerns of the protesters, but said overall the company had "received tremendous support from people on the East Coast."[44] During May 2013, Johnson maintained the charm offensive telling a *Hawke's Bay Today* reporter the company was "excited" about the preliminary findings from its Dannevirke test well and that he was keen to "remove the mystery" surrounding oil and gas development.[45] Once again he claimed TAG had received support from "a majority of people." He assured the public the company had "no plans" for thousands of fracking events and there would not be hundreds of wells dotted across the landscape (contrary to the 2013 MBIE oil and gas study which had used TAG's own projections).

TAG kept up its PR spin about positive results from its Dannevirke well in coming months. In February 2014 Chief Executive Garth Johnson said the company was following a "methodical, well-thought-out program in a frontier oil and gas basin that will take many years before we can say with any confidence if a commercial oil and gas play is possible."[46] When nothing more was heard for months, TAG shareholders and Tararua District councilors began to wonder what was going on.

Meanwhile in June 2012 the TAG lodged a further consent application in partnership with Apache Corporation, a large US company, to establish a road and drilling pad just north of Gisborne which they called Punawai-1. The JV's plans were signaled well ahead of time, sparking concern among landowners and the convening of several meetings organized by environmental activists. The question of whether the Gisborne District Council would publically notify the consent application for the road and drilling pad, allowing submissions from the public, suddenly became a big issue. Under the RMA, councils had discretion over what types of resource-use activity required notified consent. A group calling itself Frack Free Tairāwhiti (FFT) circulated a public petition calling on the Council to make all extractive applications notifiable. The petition received almost 2,000 signatures. FFT also arranged for the Environmental Defence Society (EDS), a national organization, to submit a legal opinion to the Council about the Punawai-1 application. The submission accused the JV of "gaming the system" by applying for a permit to build a road and drilling pad but not for the well itself. It was patently obvious the JV would not go to the expense of building a road and pad unless it intended drilling one or more wells. To exercise prudent resource management, EDS argued, the Council should require the companies to submit an application for the *entire project*.

Apparently the TAG/Apache partners decided it was time to go on the offensive to answer the critics and make sure the Council didn't open up the consent process for public input. Apache's senior adviser of regulatory affairs Alex Fergusson visited New Zealand in July 2012 for discussions with government officials and ministers. In an interview with the *Gisborne Herald* he argued that FFT's 2,000-person petition "should be seen by local and central government as a lack of public confidence in their ability to perform their regulatory role."[47] It wasn't necessary to notify every consent because the issues and interests in one extractive consent would be similar to others. In the meantime, he announced, TAG/Apache had deferred any "hydraulic fracturing plans" for Gisborne until the PCE's report on fracking was completed. It was a hollow gesture clearly designed to curry public favor. The JV had no plans for fracking in the district or anywhere else on the East Coast at the time.

Meanwhile, the PCE's investigation into fracking was underway and she was due to visit the East Coast for public consultations in the spring of 2012. With ongoing anti-fracking activities and media statements by green councilors, PEPANZ's David Robinson decided to speak out. In a letter to the editor of the *Gisborne Herald* in late July following publication of Apache spokesperson Fergusson's interview, he launched a thinly veiled attack on activist District Councilor Manu Caddie and his environmentally minded colleagues. He stated, "Local representatives have a responsibility to provide balanced information to residents. It's their job to put aside their own personal ideology and lay out the pros and cons of any situation to enable residents to *make an informed decision*."[48] He claimed that information about fracking and New Zealand regulations to protect the environment had not been "given out to local residents" because they went against the ideology of opponents.[49] It was an extraordinary attack on democratically elected local government officials by a spokesperson for the petroleum industry.

At the beginning of September 2012, a week before the PCE visited the East Coast, the industry and government launched what amounted to a coordinated promotional campaign to get out ahead of the fracking controversy. MBIE had released its paper on the economic potential of the oil and gas industry a fortnight earlier.[50] Apache's Alex Fergusson flew to Gisborne before going back to Canada and made a presentation to the Gisborne District Council on TAG/ Apache's plans for oil and gas exploration on the East Coast. The presentation was closed to the public.[51] The same week as Fergusson's visit, Pieri Munro, NZPaM's Iwi Liaison officer, and other officials met with Gisborne District councilors and iwi representatives to discuss establishing a "local stakeholders group." The aim of the group would be to discuss local concerns, commission reports, and "work toward a consensus" on oil and gas development. The following week, during the Commissioner's visit, PEPANZ's Robinson and his communications manager Deborah Mahuta-Coyle met confidentially with Gisborne District councilors, staff, and local business representatives to discuss how an oil and gas industry might develop in the region.[52]

Six months later, in March 2013 MBIE released its East Coast Oil and Gas Potential study. The report sparked considerable debate in mainstream and social media. TAG's COO Drew Cadenhead was interviewed in a *Gisborne Business Herald* feature several weeks later, and dismissed as nonsense claims the oil industry viewed New Zealand as the "Texas of the South."[53] It was an apparent attempt to dampen public fears (or hopes) of a possible boom, even though at the time the industry and expert consultants were touting the huge potential of the region to prospective investors. It turned out the *Herald*'s feature story was timed to coincide with Gisborne District Council's June 2013 announcement that it had received a second resource consent application from TAG (Apache had withdrawn from the JV), this time to drill a well at its Punawai-1 site. In July 2013, despite FFT's ongoing lobbying and numerous letters to the editor from the public, the Gisborne District Council announced it would not be inviting public submissions on TAG's second Punawai-1 consent application. It had decided to turn over assessment of such applications to council technical staff.

After consulting the Environmental Defence Society, FFT released a statement criticizing the Council's decision as going against the wishes of the local community:

> Oil exploration is new to our area and the council has discretion to notify if it wishes. The Environmental Defence Society's [earlier] opinion expressed the view that bundling consents was more consistent with the idea of sustainable management than splitting them up, but Council continues to allow the practice even though it could require all consents to be sought together. We are also critical of the applicants. They could ask for the applications to be notified if they really took their social license to operate seriously. But they are . . . just taking the easiest route to suit their commercial interests.[54]

Nevertheless, Council staff eventually concluded the environmental effects of TAG's exploratory well application would be "less than minor," in RMA parlance. When the issue was reported back to the Council, a majority of councilors voted to grant TAG resource consent for their exploratory well. Six months later, the Parliamentary Commissioner for the Environment in her final report on fracking and the country's regulatory regime strongly recommended that local councils make all oil and gas resource consent applications publically notifiable in light of the potential risks to the environment and the high level of community concern over fracking.[55]

In August 2013, with rumors circulating after the Council's consent decision about protestors getting organized and a possible hīkoi (protest march), the Gisborne Chamber of Commerce invited PEPANZ's David Robinson to speak about the benefits of the oil and gas industry at an invitation-only gathering. Embarrassingly for the Chamber, the confidential email to members was circulated widely to community groups and Māori organizations forcing the Chamber to cancel the meeting.[56]

At the end of March 2014, the Gisborne District Council granted TAG non-notified consent to drill another exploratory well called Waitangi Valley-1 not far from their Punawai-1 well site (which it had postponed drilling). Drilling at Waitangi Valley-1 began in late July, and a few weeks later COO Drew Cadenhead reported there had been no incidents or "complaints from local stakeholders." In truth the landowner who granted access expressed concern to a District councilor over the proximity of TAG's well site to their main water source and the fact they were powerless to do anything about it.[57]

From this point, things began to go wrong for TAG on the East Coast. In mid-September the company announced that it was shutting down operations at Waitangi Valley-1 and capping the well. CEO Garth Johnson explained the drilling had encountered extreme hydrocarbon zone pressures at shallow depth, which was "unprecedented" in the industry. International drilling experts had advised the operations be shut down for safety reasons.[58] The company told shareholders and the media it would abandon the well "for the time being" and review its program for the coming year. However, the company remained "excited about our East Coast prospects."[59]

Three months later in December 2014, with international oil and prices dropping, TAG's partner East-West Petroleum announced it was withdrawing from another permit south of Gisborne. In January 2015 TAG announced a new work program concentrating on its existing Taranaki sites in light of slumping world oil prices. CEO Garth Johnson said the company remained "keen on New Zealand" and still retained long-term goals for the East Coast.[60] In early February Johnson and Cadenhead resigned. The company's media statement said it was time for a change at the executive level "to make way for the required expertise to execute on TAG's next phase of growth" which, to some shareholders, looked more like retrenchment.[61] In September

TAG announced it was relinquishing its two remaining East Coast permits, although new country manager Max Murray indicated the company still believed in the potential of region and looked forward to returning. Right to the end, the company was selling the East Coast.

Harnessing MSM and Social Media

We have already noted examples of how key players in the New Zealand petroleum promotion network took advantage of close relations with mainstream media to manage criticism and put out positive messages about the industry. Some of these were ad hoc practices similar to those used by other industries and PR companies such as "capturing" favored reporters by giving them access to senior executives, providing them insider information, comping them to attend conferences, and inviting them to special events. Business news staff at *Otago Daily Times* for instance enjoyed special relations with the petroleum industry which they were able to use to present detailed, "balanced" coverage of the industry's activities and the controversies arising from them, while at the same time collaborating in a private subscription news service for the industry.

A favorite tactic was monitoring MSM and working behind the scenes with editors to present positive stories about the benefits of oil and gas development. And, of course, coordinating responses to critical opinion pieces and letters to generate public debate. The *Gisborne Herald*'s June 2013 quarterly *Business Herald* coverage of the industry and extensive interviews with TAG Oil executives to "put their side of the story" was a case in point. An editorial two days earlier supported TAG's efforts to correct "misinformation" about the industry and its potential impacts, praised the company's efforts to meet and communicate with rural residents and iwi, and called on people to read the upcoming feature with an "open mind."[62]

The *Gisborne Herald* editor's efforts to promote and defend the petroleum industry became more aggressive during in 2014 and 2015 as opponents of the industry used the paper to argue against oil and gas exploration and highlight climate change concerns. In October 2015, for instance, local environmental activists lobbied the Gisborne District Council to decline support for offshore exploration in a new government block offer. A representative of GNS Science (who turned out not to be a scientist but the organization's communications manager) submitted a letter to the paper assuring the public that offshore drilling was low risk. Several people wrote to the paper criticizing the GNS letter and questioning the organization's scientific neutrality. When GNS declined further reply, the editor leapt to their defense in an editorial. The following month, the editor published the results of a scientifically invalid survey the paper had undertaken which he claimed showed a majority of residents supported oil and gas exploration. He accused critics of fabricating facts and exaggerating the risks of drilling. A fortnight later, on the eve of the Paris COP21 Climate Summit, he launched another salvo claiming (in line with PEPANZ's promotional rhetoric) that transition to a low-carbon economy would take a long time, that natural gas was a transition fuel and that the Gisborne district could have a lot of it. For good measure, he suggested greater use of natural gas would help save the Arctic![63]

TAG Oil experimented briefly with the tactic of hiring a well-known media person as their public front person, as several other companies had done. In mid-2014 after the Waitangi Valley-1 well was spudded in and drilling had commenced, TAG contracted a Wellington PR firm to arrange for one of their directors Chris Wikaira (a Māori) to serve as TAG's "company spokesperson." No doubt TAG considered Wikaira's appointment an astute move since 43 percent of the Gisborne district was Māori. Wikaira had a background in political journalism, had been Radio New Zealand's Senior Māori Issues Correspondent for years and continued to appear on Radio New Zealand's *The Panel*. He had established himself in a position similar to ex-journalists Linda Clark, Rachel Smalley (who served as MC for the 2015 Advantage NZ Summit), Allan Seay (who became Anadarko's New Zealand spokesperson), and others. The key to the tactic was to acquire someone who not only had media experience but a reputation as an objective, knowledgeable, and trustworthy professional journalist. Like "good will," companies were in effect buying the public trust and confidence embodied in such a nationally known front person.

In addition to the kinds of informal campaigns noted in the last chapter, social media were a way for New Zealand pro-oil advocates (the "flakkers") to communicate with like-minded individuals locally and overseas to share information and tactics regarding government policies, political developments, and anti-petroleum activists. TAG Oil's shareholder chatroom on the investor website www.stockhouse.com is a case in point. When left-wing Gisborne District councilor Manu Caddie gave a presentation at a 2013 Oil and Gas Exploration Symposium in Hastings criticizing government policy and lack of public consultation over oil and gas development, several TAG shareholders and industry advocates exchanged views on what Caddie's influence was and how the "pro-fracking" community could respond to him in letters to the editor.

Pseudo-Consultation, Co-optation, and Divide and Conquer

PEPANZ and NZPaM have put considerable effort into engaging with communities and consulting with Māori. They have also supported and helped disseminate to the industry the best practice Māori engagement guide developed by Taranaki iwi Ngāti Ruanui. In reality, many industry engagement activities are far from the culturally appropriate processes that would afford communities or tangata whenua any real degree of influence over decisions about oil and gas development. Many "consultative" gatherings have been little more than pre-packaged company presentations followed by a Q & A session.

For example, in March 2013 Canadian company NZEC agreed to appear at a hui at Kahungunu Marae in Nuhaka, the tūrangawaewae (place to stand) of Te Iwi o Rakaipaaka to discuss their plans for exploration in the area. The panel was comprised of NZEC's Māori land access manager and a Māori "cultural auditor" contracted from Ngāti Ruanui in Taranaki, a consulting geologist, and two staff from the seismic exploration contractors. There were no company "chiefs." The hui involved the company representatives talking at the assembled audience of around fifty Māori and Pākehā (European New Zealanders) for almost two hours. When it came time for questions, several people expressed anger at being fobbed off with technical talk and assurances of how culturally sensitive NZEC was. A kuia (respected elder) said all the talk had been about exploration but everyone knew what came after that. She had recently visited First Nations people in Canada who'd been fighting oil companies for years. Tangata whenua, she insisted, should come up with an action plan to stop the drilling. At that point, the company's land access manager intervened as he had done repeatedly during the discussion, saying it was important to "understand the permitting process." The Crown had consulted on these permits and given its approval for NZEC to proceed. "So it's not about 'stopping' but how do we *engage and work together*. It's not about problems but *what are the 'opportunities.'*"[64] It was an accurate if patronizing summation of the current situation under current legislation. New Zealand communities and iwi, as we'll see in the final chapter, do not have a "no development" option in regard to expansion of the petroleum industry. A few local councils and iwi have however managed to get the Minister of Energy and Resources to alter block permit boundaries which encroached on environmentally sensitive areas or tapu (sacred, protected) sites.

Under the Crown Minerals Act, exploration permits require companies to consult with tangata whenua about company plans and report on that engagement to local councils and NZPaM. Exploration and production companies typically contract out such work. As a result, Māori end up dealing with "consultation specialists" and technicians rather than company executives who are the decision-makers responsible for company policy and operations. When the TAG/Apache JV began planning in 2011 for their first exploratory well near Gisborne, they contracted an engineering and environmental planning firm based in New Plymouth to engage with local iwi Ngā Ariki Kaiputahi (NAK) and Te Aitanga-a-Māhaki, and hired Wellington lawyer Andrew Beatson to report on the process. There was a contentious history between the two iwi which, judging by subsequent developments, neither the consultants nor Beatson seem to have been aware of.[65] An initial hui in 2012 held at Ngā Ariki Kaiputahi's Mangatu marae—primarily a technical presentation by the consultants followed by a Q&A session—was attended by "the iwi, hapū members, and local residents" according to Beatson. From his report, it is not clear how many people were in attendance and particularly how many were there from neighboring iwi Te Aitanga-a-Māhaki. He emphasized that company representatives were "formally welcomed" onto the marae. This was not special treatment; the Māori pōwhiri (welcome) is customary protocol for receiving guests. Beatson may have simply hoped to impress District Council staffers unfamiliar with Māori tikanga (customs).

In January 2012, the JV partners made a public presentation to the Gisborne District Council which leaders of the two iwi attended. This was followed by a series of meetings between representatives of the two Māori groups and JV representatives (i.e., their Taranaki consultants). Beatson adopted the practice of using the terms "iwi" and "hapū" to refer to participants in email/phone communications and meetings, evidently to convey the impression of widespread tribal involvement. In reality these exchanges seldom involved more than JV representatives and two iwi leaders.

Relations between the JV and iwi were further complicated when Ngā Ariki Kaiputahi adopted an Iwi/Hapū Management Plan at a tribal hui in 2012 and formally submitted it to the Gisborne District Council. The Plan contained a strongly worded environmental section with specific protection measures and objectives. A group calling themselves the NAK Iwi Management Team (iwi environmental activists who had prepared the Plan) began raising questions about the environmental and cultural impacts of the JV's proposed exploration program, and insisted on taking part in discussions. TAG consultants dodged the request for most of 2012 and worked with the two iwi leaders to progress the earthworks consent for their Punawai-1 well, which Council approved in September. The NAK Iwi Management team demanded to be consulted over the next stage application for drilling a well, and suggested TAG establish a Māori unit to consult properly. TAG ignored the request, preferring to leave the matter to their contracted community engagement specialists, though they did send the NAK team a copy of the consent application.[66]

In May 2013 the NAK Iwi Management team informed TAG's consultants they were unable to fully assess the potential impacts of the company's next stage plans on the iwi, and would oppose the drilling consent application "unless certain preconditions to further engagement were met." The team wanted time to examine the well consent application in light of the NAK Iwi/hapū Management Plan, and either negotiate changes with TAG or object to the application.[67] In his report to the Gisborne District Council, Beatson intimated that the NAK Iwi Management team was a breakaway splinter group and their demands were not in accord with Environment Court case law regarding what constituted "proper two-way consultation." TAG had established working relations with the two iwi leaders, done its best to the accommodate tribal activists, and intended carrying on with their exploration plans as they were entitled to do under current legislation. They weren't about to go back to iwi members on every aspect of the project, much less placate a persistent band of "stirrers."

District Council staff used Beatson's report to justify their recommendation not to publically notify TAG's Punawai-1 consent application. The Council's chief executive said they had received "proof from TAG that it had *conducted meaningful consultation with Māori.*"[68] The "Cultural Assessment" section of the staff report[69] reiterated the company had undertaken "extensive consultation" with iwi *and* the NAK Iwi Management team. The Council voted to approve TAG's resource consent in mid-August. A Ngā Ariki Kaiputahi spokesperson interviewed on national radio said the iwi was upset at not being properly consulted, and would consider an appeal and protest action.

LOCAL GOVERNMENT CHEERLEADING
AND GATEKEEPING

Industry and central government efforts to sell the East Coast on oil and gas development received mixed support from East Coast councilors, mayors, and economic development agencies. For councilors who favored *any* kind of development and thus were pro-oil, the trick was to stick to the rhetoric about the benefits, safety, and environmental responsibility of the industry while avoiding being drawn into public debate over the potential harmful effects. Local councils for the most part managed not to have to discuss such issues on their meeting agendas, though it did crop up during long-term planning exercises. Calls by critics for a comprehensive cost/benefit analysis before an oil and gas industry was allowed to be established were ignored.

In January 2012 (as noted earlier) the Gisborne District Council heard a public presentation by Alex Fergusson, Apache chief advisor for regulatory affairs, about the potential for oil and gas development in the Gisborne region. The council room was packed with staff and members of the public, including a number of anti-fracking activists and iwi representatives. Fergusson provided a brief background on the Apache/TAG partnership, discussed the stages of oil and gas exploration and production which the JV planned to follow (emphasizing "community engagement" during each stage), and answered questions from councilors.[70] During the morning tea break, councilors and members of the public had a chance to meet and talk with Fergusson, and a few activists made their concerns known to him. Toward the end of the tea break, a councilor who at the time happened to chair the Council's Environmental Planning and Regulations Committee took one of Fergusson's aides aside. "Don't worry," she was overheard to say. "We'll look after you."[71]

In July 2013 the Gisborne District Council convened as the "Future Tairāwhiti" committee, which was comprised of all councilors, to discuss the Council's attitude to oil and gas development. Some councilors were strongly supportive; others were strongly opposed. They eventually adopted a staff recommendation that a strategic response to the possibility of oil and gas development be drafted, looking at the economic challenges and opportunities that such development could pose. The idea had been mooted in the Parliamentary Commissioner for Environment's final 2014 report. However, at the next full Council meeting, Mayor Meng Foon who had been absent from the earlier meeting, torpedoed the idea saying the Council should take a "balanced view" and encourage all development "as well as environmental sustainability."[72] Foon was a local businessman and support for oil and gas development had been part of his re-election platform the previous year.

Meanwhile, Mayor Lawrence Yule of Hastings (also President of Local Government New Zealand) decided to organize a public symposium in October 2013 to debate the oil and gas issue. The first part of the symposium focused on "factual" information about government regulations, oil and gas exploration processes and the reasons to encourage exploration (in other words, NZPaM's public information roadshow format). Presentations by concerned farmers and anti-fracking activists like Gisborne Council's Manu Caddie were relegated to the end of the agenda. Some in the audience criticized what appeared to them to be a "gate-keeping" agenda but the Mayor defended the arrangements, claiming the Council had gone to great lengths to present a "balanced" agenda. The underlying issue, not addressed on the night, was whether the symposium debate had any bearing at all on whether oil and gas development proceeded on the East Coast.

Six months later, LGNZ launched a campaign in the lead up to the 2014 national elections, supporting oil and gas development and calling for royalties to be shared with regions where drilling and production occurred. Gisborne's Mayor Meng Foon was in the vanguard of the campaign, having earlier gotten himself appointed to LGNZ's Royalties for the Regions committee. If there was a major oil or gas strike, he argued, regions should receive a share of the royalties to upgrade infrastructure.[73] In August LGNZ held a forum in Wellington on "The Case for a Local Share" (of royalties) which received wide media coverage. Clearly business-dominated local authorities were keen to see the expansion of the petroleum industry in their regions, with appropriate nods to environmental safeguards and effective mitigation measures.

CONCLUSION

Selling the East Coast on oil and gas development is a case study in how the National government, specifically the ministers of Economic Development and Energy, paved the way for the industry by orchestrating a regional development campaign of targeted information dissemination, coordinated ministerial PR visits, jawboning local councils, marshalling business support, and utilizing mainstream and social media to answer environmental critics. Local councils helped by encouraging public discussion while cheerleading and gatekeeping for the industry. Some of these activities happened to coincide with petroleum industry tactics of their own, while others such as pressuring local authorities and manipulating the media appear to have been occasionally coordinated (e.g., during the Parliamentary Commissioner for the Environment's public consultation visit to the East Coast).

The key to the effectiveness of this collaborative sales campaign was maintaining the myth—through MBIE studies, ministerial media statements, presentations, and PR by industry representatives—that the citizens of the East Coast could influence decisions about whether and how an oil and gas industry developed. Under current government policy settings and legislation, that was never going to happen. The trick was to get people to engage in debate and discussion, "have a look" at oil and gas, and get behind central government-led regional development plans as if these represented the will of the local citizenry.

In terms of community engagement a few attempts were made to establish "stakeholder forums" but these did not eventuate, possibly because most companies relinquished their East Coast permits at least for the time being. Government and exploration company consultation exercises were criticized as culturally flawed, perfunctory and/or manipulative. For the most part environmentalists, anti-fracking groups, and Māori activists were treated as troublesome anti-development naysayers. They were not invited to join official working groups, and seldom participated in consultative meetings with ministers or hui with petroleum representatives. More than simply ostracizing organized groups and dismissing persistent individual critics, some local councils were prepared to take industry gatekeeping to the extreme by barring public involvement in resource consents even though they had the discretion (indeed encouragement from the PCE) of opening the process to public submissions. In the case of the Gisborne District Council, the majority of councilors voted to hand over the assessment of consent applications to technical staff, turning a blind eye to TAG's disingenuous Māori consultation report which Council staff used to support their "no public notification" recommendation, and allowing the company to get away with "gaming" the resource consent process.

Government's perceived obstinacy over calls to reconsider its energy and climate change policies, coupled with industry PR spin about the industry's benefits and moves by local councils to block public participation in drilling consents fueled a growing conflict consciousness[74] among activists and concerned citizens alike. Judging by feedback on news stories, talk shows, and letters to the editor, more people seemed to believe that targeted publicity campaigns, lobbying, and even protest action may be necessary to force policy changes and a perhaps change of government.

But there was also growing frustration on the part of local councils at the government's legislative reforms, perfunctory block offer consultation process, refusal to consider sharing oil royalties, and takeover of regional economic development planning. As a result, there was an increasing imbalance in relations between institutions that had to do with national and regional development and resource management. Tensions increased between a government determined to impose its will and force the pace of oil and gas expansion, a petroleum industry keen to exploit the rumored riches under the region, and concerned citizens and environmental activists equally determined to block government and industry efforts to "transform" the economy of the East Coast.

In the next chapter we will examine how communities, environmental organizations and indigenous activists in New Zealand and elsewhere have responded to such developments.

NOTES

1. The East Coast is the area east of the North Island's central mountain range that stretches from the Wairarapa District near Wellington northward to East Cape at the top of the Gisborne District.

2. Ministry of Business, Innovation and Employment (MBIE), "East Coast Regional Economic Potential Stage 2: Economic Forecasting and Transport and Skills Implications" (New Zealand Government, April 2014).

3. MBIE, "Economic Contribution and Potential of New Zealand's Oil and Gas Industry."

4. New Zealand Institute of Economic Research (NZIER), "East Coast Oil and Gas Development Study: Economic Potential of Oil and Gas Development," Report to the Ministry of Business, Innovation and Employment (New Zealand Government, Wellington, November 2012), http://www.mbie.govt.nz/info-services/sectors -industries/natural-resources/oil-and-gas/petroleum-expert-reports/east-coast-oil-and -gas-development-study/NZIER-report-East-Coast-study-Value-oil-gas-exploration .pdf. Michael Adams, "East Coast North Island – Oil Resource Play – Development Scenario Models." Report to the Ministry of Business, Innovation and Employment (New Zealand Government, Wellington, October 2012), http://www.mbie.govt.nz/ info-services/sectors-industries/natural-resources/oil-and-gas/petroleum-expert -reports/east-coast-oil-and-gas-development-study/MARE-EC-Resource-Plays -scenarios-East-Coast.pdf.

5. MBIE, East Coast Oil and Gas Development Study, 5. Emphasis added.

6. MBIE, East Coast Oil and Gas Development Study, 5.

7. MBIE, East Coast Oil and Gas Development Study, 6.

8. MBIE, East Coast Oil and Gas Development Study, 6.

9. The MBIE authors acknowledge that in North America an unconventional well might require between a few thousand and 20,000 cubic meters of water. A conservative estimate of water demand for fracking and production on the East Coast, assuming Bakken-like geology and half the average usage reported in several US studies (see chapter 2) taking account of the New Zealand unconventional industry's low level of infrastructure, suggests total water demand under MBIE's scenario 5 could range between 5.5 million and 540 million cubic liters over the 30+ years.

10. Richard Heinberg, 2013, *Snake Oil*, 70. The North Dakota Department of Mineral Resources Monthly Production Report indicated the average output per well in 2014 was between 130-140 BOE per day, or approximately 306,000 over a well's lifetime. See the North Dakota State Government portal, https://www.dmr.nd.gov/ oilgas/mprindex.asp.

11. Sproule International, "Technical Assessment of the Undiscovered Hydrocarbon Resource Potential of Pep 38348 and 38349, Onshore East Coast Basin, New Zealand (as of May 31, 2008)," 41, accessed November 17, 2014, https://www.scribd .com/document/35360884/TOP-Sproule-Report-2008-5-31.

12. MBIE, 2013, "Terms of Reference for the East Coast Regional Economic Potential Study," MBIE-MAKO 4228738 (New Zealand Government, Wellington, May 1, 2013), Team leader, Felicity Bowen, file:///C:/Users/Terrence/Downloads/ East-Coast-Study-TOR-1-May-2013.pdf.

13. MBIE, "East Coast Economic Potential Final Report: Stage 1" (New Zealand Government, Wellington, April 2014), 106, emphasis added.

14. MBIE, 2014b, "East Coast Economic Potential Final Report: Stage 2," 22.

15. MBIE, 2014, "East Coast Economic Potential Final Report: Stage 1," 3.

16. MBIE, 2014, "East Coast Economic Potential Final Report: Stage 2," 29.

17. A New Zealand Energy Corporation executive addressing the Hawke's Bay Regional Council in 2013 predicted there could be thousands of wells on the East Coast by mid-century.

18. MBIE, "East Coast Economic Potential Final Report: Stage 2," 46 ff.

19. MBIE, "East Coast Economic Potential Final Report: Stage 2, 47.

20. Personal communication from a Gisborne District Council councilor, June 8, 2014.

21. "Oil Well Search Plans," *Gisborne Herald*, July 16, 2014, accessed September 17, 2014.

22. Gavin Maclean, "Oil, Gas and a Lot of Hot Air, "Opinion, *Gisborne Herald*, July 19, 2014, accessed September 17, 2014.

23. Hon Simon Bridges, "Important to Have Facts on Table," *Gisborne Herald*, July 19, 2014, accessed September 17, 2014.

24. "MBIE Spend on Media Monitoring – Context," *Scoop News*, October 23, 2015, accessed October 24, 2015.

25. "MBIE Spends Up Large on Media Tracking Services," *Television New Zealand*, October 23, 2015, accessed October 24, 2015.

26. Jawboning is a political and economic term used especially in the United States, alluding to the Old Testament account of Samson using the jawbone of an ass to slay his enemies. It refers to unofficial pressure or moral suasion based on an implied threat (e.g., government regulation, economic penalties, reputational damage).

27. Discussion with Cr Andy Cranston, August 21, 2012. Also Andy Cranston, "Meeting with Ministers," Report 12/452, Gisborne District Council, 21 August, 2012, https://www.DOCS_n265120_v1_Council_Late_Item_12-452_Report_Meeting _of_Ministers.doc1/2/.

28. Cranston, "Meeting with Ministers," emphasis added.

29. Hon Simon Bridges, "Committed to Upholding Oil and Gas Safety Record," *Gisborne Herald*, November 9, 2013, accessed August 21, 2015, emphasis added.

30. Pauline Doyle, "Minister Misrepresents Interim Fracking Report," letter to the editor, *Gisborne Herald*, November 16, 2013, accessed September 17, 2014. Doyle was a spokesperson for the lobby group Guardians of the Aquifer.

31. Communication from a confidential Gisborne business community informant, November 20, 2013.

32. "Have a Look at Oil and Gas," *Gisborne Herald*, December 7, 2013, accessed October 13, 2014.

33. "Taranaki's Economy Lagging Behind Regional Counterparts," *Taranaki Daily News*, August 31, 2015, accessed September 5, 2015.

34. "Region Must Grab Its Opportunities: Joyce," *Gisborne Herald*, December 9, 2013, accessed October 13, 2014. Sir Apirana Mahuika, the leader of Ngāti Porou, had recently announced that the iwi took its mana whenua responsibilities seriously and was opposed to oil and gas exploration on the East Coast.

35. "Exploration Permits 'Show Disregard' for Ngāti Porou." *Gisborne Herald*, December 7, 2013, accessed September 17, 2014.

36. "PM: Mining, Oil Huge Potential for Region," *Gisborne Herald*, August 30, 2014, accessed October 14, 2014.

37. "Region Dead Last in Economic Rankings," *Gisborne Herald*, October 17, 2015, accessed October 18, 2015.

38. "Councillors 'Kept in Dark' over Economic Plan," *Gisborne Herald*, May 12, 2016, accessed May 20, 2016.

39. "TAG Oil: Canadian Owned, New Zealand Run," *Exploration World*, September 2013, www.explorationworld.com.

40. "TAG Oil Welcomes Drilling Consents," *Scoop News*, March 18, 2013, accessed November 10, 2014, emphasis added.

41. "Tararua Farmers Stage Fracking Protest," *Hawke's Bay Today*, April 8, 2013, accessed September 18, 2014.

42. "Protesters Fear for HB Water," *Hawke's Bay Today*, April 9, 2013, accessed September 18, 2014.

43. Anyone who was able to locate the report on the Taranaki Regional Council's website and took time to read the fine print would have learned that two of nine test holes found high levels of methane and unnatural chemicals. An independent reviewer suggested it was only possible to speculate about the causes since the monitoring was faulty; no control tests had been carried out before drilling operations began.

44. Results of a national survey presented by PEPANZ's David Robinson to the 2012 New Zealand Impact Assessment conference showed that East Coast interviewees were least favorable to the oil and gas industry.

45. "TAG Keen to Remove Oil Mystery," *Hawke's Bay Today*, May 28, 2013.

46. "TAG Oil Firm May Stop Short of Fracking," *Hawke's Bay Today*, February 24, 2014, accessed August 19, 2014.

47. "Petition Shows Lack Of Confidence in Regulators: Apache," *Gisborne Herald*, July 3, 2012, accessed September 17, 2014.

48. David Robinson, "All or Nothing Approach No Good for Communities," *Gisborne Herald*, July 31, 2012, accessed September 9, 2014, emphasis added.

49. This is not a council function under the Local Government Act. To do so would leave the council open to accusations of bias.

50. MBIE, "Economic Contribution and Potential of New Zealand's Oil and Gas Industry," Occasional paper 12/07 (New Zealand Government, Wellington, August 2012).

51. Personal communication from a Gisborne District Councilor, September 16, 2012.

52. Personal communication with a Gisborne District Councilor, September 16, 2012.

53. "Business Herald Quarterly: Oil and Gas Industry Feature," *Gisborne Herald*, June 10, 2013. The phrase "Texas of the South" appeared on TAG Oil's website in a 28-page paper for investors titled "TAG Oil: 2011 – A Record Year" in December 2011 ("NZ 'Likely Texas of the South,'" *Stuff News*, January 15, 2012, accessed May 27, 2015). The phrase was also used by a TAG executive in a subsequent 2012 Radio New Zealand interview.

54. Email communication with a member of Frackfree Tairāwhiti, August 13, 2013.

55. Parliamentary Commissioner for the Environment, "Drilling for Oil and Gas in New Zealand: Environmental Oversight and Regulation" (New Zealand Government, Wellington, June 2014), 68 ff.

56. In early February 2014 Robinson was invited back to speak to an invitation-only Chamber meeting [information provided in an interview with a senior Gisborne District Council official]. However, on February 20, NZEC announced he had resigned from PEPANZ to join the company.

57. "TAG Oil Begins Drilling Waitangi Valley-1 Unconventional Exploration Well," *Market Watch*, July 24, 2014, accessed July 31, 2014. Information from personal communication with the District councilor involved, June 6, 2014.

58. "TAG Oil Announces Abandonment Plans at Waitangi Valley-1 and Return to Drilling of Core Production Assets in Taranaki." *Oil and Gas News*, September 16, 2014, www.youroilandgasnews.com.

59. "Tag Oil Closes Lid on Bay Well," *Hawke's Bay Today*, September 18, 2014, accessed September 19, 2014.

60. "TAG Oil's Emphasis on Low-Cost Drilling to Sustain Strong Balance Sheet," *Proactive Investors Australia*, January 8, 2015, www.proactiveinvestors.com.au. Also "Oil Firm Reviews Exploration Plans," *Hawke's Bay Today*, January 20, 2015, accessed January 24, 2015.

61. "TAG Oil Announces Executive Changes," *Market Watch*, February 11, 2015, accessed February 12, 2015. Shareholder comments on TAG's *Stockhouse.com* page were pessimistic about the changes. Six months later, a note in the national business media announced Cadenhead and Johnson had formed a company to exploit the East Coast's offshore methane hydrate deposits. Government-funded studies and oceanographic surveys into methane hydrates by GNS Science and NIWA had been underway since 2009.

62. "Being Informed a Good Start," *Gisborne Herald*.

63. "Fossil fuels Will Be with us for a Long Time to Come," *Gisborne Herald*, November 27, 2015, accessed November 28, 2015.

64. Field notes from attendance at the Nuhaka hui, March 14, 2013, emphasis added.

65. Andrew Beatson, "Summary of Consultation between TAG Oil and Iwi in Relation to Proposed Exploration Well Known as Punawai-1," Gisborne District Council, letter to Hans van Kregten, Environment and Policy Group Manager on July 8, 2013, www.gdc.govt.nz. Beatson was an experienced environmental and resource management lawyer with Wellington law firm Bell Gully. In his report he erroneously referred to Ngā Ariki Kaiputahi as a hapū (sub-tribe) and implied it was part of Te Aitanga-a-Māahiki iwi. Ngā Ariki Kaiputahi traces back to a separate apical ancestor and claims mana whenua over land adjacent to and in some instances overlapping with Te Aitanga-a-Māhiki. See http://www.gdc.govt.nz/assets/Files/Iwi-Plans/Hapū-Iwi-Management-Plan-of-Ngā-Ariki-Kaiputahi.pdf.

66. Information from an interview with a member of the NAK Iwi Management team on March 4, 2014. The interviewee was informed that material could be used in publication and gave consent for its use on condition of anonymity.

67. Information from an interview with a member of the NAK Iwi Management team.

68. "Conditions Will Apply to Oil Well," *Gisborne Herald*, August 16, 2013, accessed August 20, 2013, emphasis added.

69. Gisborne District Council, "Report 1 – District Landuse Resource Consent: Punawai-1 Exploratory Well Drilling, Testing and Associated Works" (Gisborne District Council, Gisborne, August 2013). The Council voted to defer resource consent decisions to technical staff.

70. Gisborne District Council Meeting Minutes, Agenda Item 8.11 No. 12/031, "Oil Exploration and Production," January 26, 2012.

71. Personal communication from a member of Frack-free Tairawhiti who was present at the Council meeting, January 29, 2012.

72. "Pondering Vote Against Strategic Response to Oil and Gas Issues," *Gisborne Herald*, August 8, 2014, accessed October 17, 2014.

73. "Share of Oil/Gas Earnings 'Should Come to Regions,'" *Gisborne Herald*, April 1, 2014, accessed April 2, 2014.

74. Saul Alinsky, *Rules for Radicals* (London: Vintage Books, 1971/1989).

Chapter Six

Community and Indigenous Responses to Oil and Gas Development

THE EFFECTS OF EXTRACTIVE BOOMS ON COMMUNITIES AND INDIGENOUS PEOPLES

The social, economic, and environmental effects of oil or gas development on a community can be significant depending on the size of a given field and the scale of operations. Added to these are the wider structural influences of government policies and political agendas, and the economic and political influence Big Oil is able to wield to push through its projects (see chapter 1). Petroleum mega-corporations have achieved their power in part through deregulation of local economies, free trade agreements, and undermining the ability of communities to control their own destiny.[1] As a consequence, local communities and first nations with their own distinctive history, culture, economy, and long-standing social relationships are increasingly under threat from global corporate interests mediated by compliant states.

In chapter 2 we examined evidence of the negative effects that unconventional oil and gas production can have on human health and the environment. The boomtown (or social disruption) model also proved useful in highlighting the boom-and-bust nature of such extractive industries, their impact particularly on small communities, and the fact that booms typically reach a tipping point triggering sudden social and economic changes that can overwhelm communities and municipal governments. Recent research on the global boom in unconventional oil and gas has found that the boomtown model is still useful, though the picture is more complicated than first thought.

Community responses to the prospect of nearby oil and gas developments tend to go through phases. Bebbington observed from a number of South American case studies that these phases were related to the phases of an extractive project itself. Key stakeholders may disagree strongly when a project is in the planning stage about whether the economic, physical, and social effects will be good or bad for the community. Then once the project is underway, they disagree over how to interpret and evaluate the changes taking place.

> Crucially, these disagreements are not only *between* communities, NGOs, states and companies, but also exist *among* community members, *among* different parts of the state, *among* NGOs and even (if less so) *among* company staff.[2]

Other studies suggest most of the local population may initially adopt a wait and see attitude, although people with some knowledge of the controversies surrounding the industry may be against such a project from the outset. Public attitudes begin to crystalize into different "camps" as industrial activity picks up, jobs are created, and certain business sectors benefit while the negative impacts begin to become more apparent. Communities and indigenous peoples are seldom unanimous in their responses to oil and gas development, and responses can change over time. McAdam and Boudet[3] found in a comparative study of twenty US communities facing new energy projects that the overall level of opposition to the projects was low at least initially. However, citizen concerns over environmental impacts had a significant influence on altering the projects. Apparently protection of the environment was an issue that united more people regardless of their differences over the benefits and costs of an extractive project.

All of this is to say that it would be inappropriate to treat communities impacted by extractive developments as though they were monolithic entities. This certainly becomes apparent in the case studies that feature in this chapter. The question is, given the potential for different phases of response and internal conflicts, what have communities and indigenous groups done to understand and respond to the external forces impinging on them and retain a degree of control over their own destiny?

This chapter focuses particularly on some of the practices that local environmental organizations, anti-petroleum groups, and indigenous activists (in New Zealand's case, Māori people) have adopted to challenge government fossil fuel policies and block or modify petroleum company projects. Some of these strategies and tactics are home grown but most are similar to, or even borrowed from, overseas anti-fossil fuel movements and community action groups. Research suggests that community groups, environmental organizations, and indigenous nations are more effective in resisting extractive developments or minimizing their harm when they have wider communication networks and allies, and understand the national and international forces bearing on their struggle which they must also contend with. The chapter includes with two case studies of locally organized opposition to unconventional oil and gas projects, one in New Zealand's petroleum region of Taranaki and the other on the East Coast where preliminary oil and gas development has begun to occur.

COMMUNITY AND INDIGENOUS STRATEGIES IN RESPONSE TO PETROLEUM DEVELOPMENT

Recent research[4] suggests there are some typical actions that concerned community groups, environmental organizations, and indigenous activists have adopted to understand, prepare, resist, and/or exert a degree of influence when the oil and gas industry arrives. The actions frequently mentioned in the literature include:

1. Grassroots research, sometimes referred to as "citizen science"
2. Public awareness-raising, information dissemination and education
3. Engaging with and influencing government institutions
4. Organization, multi-level networking and alliance-building
5. Confronting oil and gas corporations and exposing their strategies
6. Indigenous self-determined development and resistance

Let's consider a few examples of these activities and how they have been employed by community groups and indigenous activists in New Zealand.

Citizen Science: Community-Driven Research

"Citizen science" refers to scientific work carried out by members of the general public, often in collaboration with or under the direction of professional scientists and scientific institutions. In recent years, there has been growing interest among environmental organizations, community anti-oil groups, and indigenous activists in planning and undertaking their own research, monitoring, and evaluating extractive projects in collaboration with scientific advisors and mentors. Terms like "citizen monitoring" and "participatory action research" reflect the community-driven nature of such activities.

Community-initiated information gathering and action research has become a fairly common practice where communities have been affected by rapid and protracted unconventional oil and gas development. As Widener discovered,[5] communities and local councils often find themselves having to respond without adequate forewarning of what is about to occur. There may be a dearth of information or, on the other hand, councils and citizens may be overwhelmed with reports, studies, government, and corporate media releases, new regulations, and changes to consent processes.[6] It is difficult for concerned citizens to engage in their own literature reviews, evidence-gathering, analysis, and monitoring because they usually lack the resources, expertise, and/or time. It falls to a few dedicated volunteers to lead the fight. In his Marcellus Shale study Wilber[7] referred to such people as "accidental activists": concerned residents alarmed by what's happening, angry at being left out of planning decisions or conned by companies, driven to learn more about the petroleum industry, scanning the Internet for research reports and information about what other communities are doing, searching land records and consent agreements, developing maps of drilling activity, monitoring impacts, and helping organize responses.

In other words, community-based citizen science is about more than just information-gathering to document harmful effects of drilling operations. Company representatives and government advocates of oil and gas development frequently accuse community anti-petroleum activists of being ill-informed about the industry and evidence of its safe operations. At the same time, the petroleum industry is well-practiced in disparaging competing evidence, impugning the reputation of independent scientists, and avoiding responsibility for the long-term cumulative effects of their operations on communities and the environment. To address these issues, citizen science frequently expands in scope beyond the local scene to include compiling national and international research findings, and analyzing industry misinformation tactics and promotional propaganda.

In the case of the Union County, South Dakota, oil refinery discussed in chapter 3, two leading anti-refinery community organizations engaged in citizen action science to analyze online data and compile disaster portraits of other refinery towns.[8] They used census data and other sources to compare counties across the United States and show that those where refineries were located were associated with relatively high poverty rates, an unstable employment environment with a high turnover of temporary or contract workers, and severe housing problems. As Widener notes,

> These organizations, empowered by online information, conducted their own grassroots cost/benefit analysis, and by appropriating what had been an industry-friendly tool found evidence to oppose the refinery. Through their research activism, they identified a pattern of social and environmental disasters associated with the petrochemical industry—and made this information easily accessible. . . . Making data easily and publicly available is in stark contrast to the secrecy and lack of transparency that is more common in industrial decisions, projects and proposals.[9]

Similarly, over the past decade as Colorado experienced an unconventional oil and gas boom, citizen groups began systematically examining the issues, searching for factual information to counter petroleum industry misinformation and PR spin, advocating for responsible policy actions, and in some instances pushing for moratoria on fracking. In 2013 in spite of an expensive gas industry media campaign, four Colorado communities passed anti-fracking referenda. According to an organizer for Citizens for a Healthy Fort Collins, "People understand half-truths when they see them. They did their research."[10] Gretchen Goldman, a scientist who reported on the Colorado community campaigns for Union of Concerned Scientists, observed that community-initiated research gave citizen groups access to the scientific and other information they needed to inform their decisions and achieve better outcomes for their community.

Concerned groups of citizens, NGOs, and academics have developed a variety of tools and methods for gathering evidence and evaluating the effects of unconventional oil and gas development in their area. These tools include conventional social impact assessment, "integrated assessment," and "holistic cost/benefit analysis." The latter approach attempts to compare and evaluate all the potential or actual costs and benefits of a development project or new industry, not just the economic impacts. For example, Kinnaman[11] examined company data and research reports about shale gas production in the Marcellus Shale and found the benefits and costs were not evenly distributed. Companies received most of the benefits and avoided most of the costs. Some externality costs went unrecognized or unmeasured and so were inadvertently left for communities or the environment to pay.

Such tools, unlike the limited environmental impact assessments that companies are required to submit for resource consents, are rarely applied in New Zealand to evaluate the potential impacts of petroleum development. However, ecologists Kyleisha Foote and Mike Joy of Massey University demonstrated the potential value of such a tool for local authorities and community activist groups. They carried out a cost/benefit analysis of the dairy industry and found that once the cost of irrigation, fertilizer, and environmental impacts were included, dairying may have cost the country billions of dollars more than it earned in 2012.[12] Following release of the study, at least one regional council indicated interest in applying a similar cost/benefit approach in considering whether to allow the establishment of an oil and gas industry. Several attempts have been made by Māori scholars and academic teams to model a holistic cost/benefit approach to assessing prospective economic development projects using a Māori cultural framework.[13] There has been limited discussion to date between New Zealand academics and local communities about applying such models to understanding the true costs of oil and gas development. However, Professor Avner Vengosh of Duke University's Nicholas School of the Environment was invited to New Zealand in early 2012 to address several hui on the impacts of oil and gas development. At the Gisborne hui, he encouraged residents to get involved in independent monitoring of drilling and production operations if petroleum projects were given the green light. He urged people concerned about the prospect of unconventional petroleum drilling and fracking to pressure universities to use their know-how and expertise to ensure the public was properly informed and involved in the decision-making process before any drilling was approved.[14]

Current New Zealand legislation[15] does not require councils to undertake a comprehensive cost-benefit analysis of a particular development project. Neither are regional economic development agencies required to carry out such analyses. If concerned citizens groups, NGOs, or interested academics wish to provide evidence-based critiques of resource consent applications or development proposals, they have to undertake their own data-gathering and evaluation exercise. This can be a challenge, because most local residents lack the time and money to get involved in this kind of voluntary activity. Furthermore, in order to have their submission treated seriously citizen-researchers have to generate popular demand for such an initiative and/or partner with professional scientists to create a mandate for such analyses to be considered in local authority planning processes.

One way of doing this is by establishing what Widener terms "legitimate physical and virtual (online) spaces" for debate, experimentation, negotiation, and decision-making.[16] As examples, she cites whole-of-community participatory budgeting in Brazil, a citizens alliance challenging the impacts of oil and gas exploration in New Mexico, and a community advisory council which provided feedback and input on a proposed refinery in Minnesota. In Gisborne Te Waananga o Aotearoa (Māori learning center) has established a three-year course leading to a bachelor of Māori advancement (environment). During the course, students discuss what a holistic cost/benefit approach would look like based on a Māori worldview and cultural values. They are applying this modeling to assist local Māori communities with their development planning, and may adapt it to provide input into broader regional issues affecting Māori like freshwater management and oil and gas development.[17]

Public Awareness-Raising, Information Dissemination, and Education

As noted earlier, research on extractive boomtowns confirms that communities are often caught unaware by the rapid growth of oil and gas developments in their area because of a lack of information about the industry and its effects. Perversely, as noted earlier, petroleum exploration and production companies often blame concerned citizens groups for being ill-informed about oil and gas operations and misinformed about how "safe" and regulated they are, even as they use their own misinformation campaigns to cast doubt on evidence-based research and monitoring reports. As TAG Oil's Chief Operating Officer Drew Cadenhead commented sarcastically during an online conference call to shareholders about rural land owner concerns over the company's exploratory well near Dannevirke on New Zealand's East Coast: "Drilling is new to these people. I think they were expecting wooden derricks and a Spindletop blowout or something."[18] He dismissed local landowner concerns about their exploratory well, saying they were simply ill-informed about industry operations.

Public information dissemination and awareness-raising is one of the first activities local anti-oil/fracking groups engage in because local petroleum projects and responses to them may not be covered in the national media, at least initially. They may only be given cursory treatment by the local media particularly if editors and owners are in the "growth at all costs" camp. Or if the state is running a pro-oil propaganda campaign in support of expanded oil and gas development, as has been the case in New Zealand.

For these reasons, community activist groups have recognized the importance of disseminating the information and evidence they have gathered by establishing websites; organizing public meetings, symposia, and debates; engaging with schools; writing to their local paper; and engaging in social media campaigns. For instance, in the debate over a proposed refinery in Elk River, South Dakota, Widener observed that refinery opponents disseminated information and expressed their concerns though individual blogs, group websites, public forums, and town hall meetings, all designed to "generate discussion in order to influence opinion, policy and practices."[19] Similarly, both McGraw and Wilber described how opponents of Marcellus Shale gas development convened informal discussion groups, published material on their own websites, and organized town meetings and symposia with invited scientists, academic researchers, and university extension staff to provide authoritative information and answer gas company and government misinformation.[20] In North America a plethora of national environmental organizations, non-profit research institutes, and environmental advocacy websites are available for community groups to tap into for information and news about what other communities are doing.[21]

Most community anti-fossil fuel groups that persist over time eventually recognize the importance of not only spreading information and raising awareness of issues, but analyzing, critiquing, and responding to government propaganda and industry greenwash. In the Marcellus Shale, for example, Bateman[22] tells how a pioneering activist group called Damascus Citizens for Sustainability came to realize it had to confront the corporate and political interests stacked against them and the vast amount of money at stake. According to David Carullo, one of the group's leaders, "what it is we're doing here is trying to dismantle the whole propaganda machine that the industry is involved in. For example, 'natural gas is the bridge to the future.'" Similarly in her comparative study of environmental contestation regarding fossil fuel development in New Zealand and Canada, Kidner[23] noted how communities and indigenous groups had to contend with what she termed "discourses of appropriation" by the state and Big Oil (e.g., talk of environmental responsibility, a social license, and stakeholder partnerships). Some local groups and national movements became adept at dealing with such discourses through what Kidner calls "re-entextualisation"—analyzing, exposing, and turning such industry and state propaganda messages on their head.[24]

At a national level in New Zealand, a number of networks and movements have arisen committed to confronting the petroleum industry and championing transition to a clean-energy future. Some prominent examples are Kiwis Connect for a Positive Future, [25] Generation Zero (NZ),[26] Wise Response,[27] the Get Free Movement,[28] and 350 Aotearoa, which is allied with Bill McKibben's international divestment movement 350.org. Some environmental NGOs such as Greenpeace, Forest and Bird, the World Wildlife Fund, ECO, and Climate Justice Taranaki have prepared investigative reports, organized national anti-oil campaigns, and served as contact points for community groups to obtain information, contacts, and support with local campaigns, government submissions, organizing public symposia, and social media campaigns.[29] The cumulative effect of the activities is that environmental issues have been constantly in the media and discussed in the public arena.

Organizing, Networking, and Planning

Concerned citizens are often stimulated to get organized and take collective action by a particular catalyst such as the announcement of a local exploration project or release of a major government policy regarding petroleum development. As we have seen in regard to the Marcellus Shale gas boom and case studies from South Dakota, Colorado, Texas, and South America, it is typical for such organizations to evolve from informal networks that coalesce around one or two articulate, determined leaders. Case studies of how community resistance groups respond to unconventional and offshore oil and gas development demonstrate the value of connecting with networks outside the community and establishing multi-level alliances, particularly where communities are under-resourced, isolated, and/or disadvantaged. Efforts by several communities in Colorado to ban fracking and control local development spawned the National Community Rights Network.[30] Networking and coalition-building sometimes generates its own set of problems, particularly where the contacts, offers of assistance, and advice are initiated from outside the community or even outside the country. As several of the chapters in Bebbington's compilation of South American case studies reveal, unless the purposes for the collaboration, the reality of the situation on the ground, and the resistance agenda are clearly understood and agreed, such coalitions can run into significant problems.[31]

New Zealand has several long-established national environmental organizations and networks that challenge the National government's push to expand oil and gas development by organizing protest campaigns and supporting local community initiatives. One of the most high-profile of these, perhaps not surprisingly, is Greenpeace Aotearoa. This local branch of Greenpeace has reputation for networking, communicating, and supporting concerned community groups and local protests where these align with their own national and international campaign priorities.

Another prominent group is Environment and Conservation Organisations of Aotearoa New Zealand (ECO),[32] a coalition of conservation and environmental groups established in the early 1970s during the Save Lake Manapouri campaign. ECO supports communities and environment groups by providing information and advice on organizing local action, research and advocacy, and networking. One of the ECO's key concerns is fracking and deep-sea drilling. In January 2015 their national conference in Taranaki was on the theme "Taranaki's Beauty and the Beast—Community Issues with Fossil Fuels and Climate Change."[33] The conference staged a one-day blockade of a Shell Todd Energy Services gas distribution site and passed a declaration calling for a nationwide ban on fracking. The declaration was formally released to the media in August after member organizations had had a chance to consider and signed it.[34] Among other points, the declaration called on communities and landowners to "Lock the Gate," i.e., declare their land frack free or mining free and deny access to oil and gas companies. The reference was to the Lock the Gate movement in Australia, with whom ECO and several member organizations had established links in 2012 after a visit to New Zealand by Australian organizer Drew Hutton.

Fractivists Aotearoa is a more informal network that emerged following the government's first block offer and the rise of anti-fracking protests. The network seems to have grown spontaneously and there is never any security vetting of participants. The network functioned more like an on-line chat room about issues, news, and protest tactics. Rather than represent organizations, participants simply expressed their own views although many were members of recognized activist groups.

Engaging with Central and Local Government

Anti-oil/fracking groups in oil-producing states try to influence governing authorities through a range of activities including lobbying individual representatives, making submissions to government bodies or public hearings, organizing petitions and referenda, participating in local planning, and mounting election campaigns for parties and candidates sympathetic to their cause. Some politically astute groups have become adept at researching the structure, functions, and membership of the relevant government entity they are dealing with and mapping the relations between key stakeholders who can influence decisions about unconventional oil and gas policies and developments.[35]

High-profile community debates, referenda, lobbying, and protest actions can undoubtedly influence how state/provincial and even national legislatures develop policy and laws around oil and gas development. But Widener observes,[36] based on her research in South America, Southeast Asia, and the several US states, that it is a difficult struggle for local activist groups to influence high-level policy decisions because most mining and petroleum projects are imposed on communities regardless of how much debate and engagement occurs. Typically such developments are justified as being in the local and/or national interest. Communities are not in a position to say no. Given the power wielded by major corporations in collaboration with national governments, communities are at the margins of the democratic process.

Although the unconventional onshore and deep-sea oil and gas industry in New Zealand is comparatively small by international standards, expansion of exploration activities under National has seen the emergence of a number of community-based organizations such as Oil Free Otago, Oil Free Wellington, Don't Frack the Bay, Energy Watch Taranaki, and Climate Justice Taranaki (CJT). They are recognized in their communities as watch dogs, lobbyists, submitters, and campaigners regarding local government policies, hearings, resource consents, and policy deliberations. CJT, for instance, is a grassroots organization that has developed expertise in researching and monitoring petroleum industry projects, making submissions on central government legislation and local government consenting and planning issues, and disseminating information about their various battles and climate justice issues via their newsletter blog and other social media. As one of the CJT leaders explained:

We need to get people to understand the linkages between social justice and climate change. They're not two things, but interconnected. Social injustice is both the cause and the result of climate change because of how the global [economic and political] system works. You have the rich ruling over the poor, and it means resources are being over-extracted leading to climate change. . . . And because of climate change, it's going to make it [poverty and the gap between rich and poor] even worse. So injustice is both the cause and the result of climate change.[37]

CJT has incorporated itself as a society for protection against vexatious damages claims by oil companies and the Taranaki Regional Council. They recently have been involved in hearings by the Environmental Protection Authority over STOS's application for a marine consent to continue operating the Maui gas field another thirty-five years. They also made submissions to the South Taranaki District Council over a plan change that would reduce the council's oversight of hazardous petroleum operations (which the industry has lobbied for), the Taranaki Regional Council's draft water plan, and on the Government's Resource Legislation Amendment Bill.

Another example of the emergence of grassroots activism in New Zealand's petroleum region is the small township of Tikorangi. Home to dairying and lifestyle blocks, Tikorangi is one of the most intensively drilled places in the country with more than ninety wells. The community has gained notoriety as a base of resistance through the efforts of a small, articulate group of citizen activists backed by a network of supporters from around the country. The group's influence stems from their persistence in pressuring local authorities and confronting companies, their effective use of communication tools,[38] and in serving as on-the-scene reporters to monitor and photograph impacts and tell the story of what's really happening to counterbalance company propaganda and government PR hype. This continuing opposition, in spite of industry claims that most residents in Taranaki support the drilling, is an embarrassment to the oil and gas industry and the government who hoped Tikorangi would be a useful test case for managing communities when land-based exploration was extended to other regions.[39]

Abbie Jury, a member of the Tikorangi residents' action group, also edits the group's newsletter blog. Jury and her colleagues fault the district and regional councils for not responding quickly enough to the rapid development of unconventional oil and gas operations in their area when the boom became apparent a decade ago. Instead of proactive planning to control the pace of drilling and mitigate the negative impacts, they claim the District Council has worked to facilitate fracking operations and side-line opponents. When there was an upsurge in drilling resource consents, the Council moved to tighten the definition of who was an "affected party" and made most resource consents non-notifiable. In spite of submissions and intense lobbying, residents feel councilors and officials choose to ignore the impacts on the community and focus on the benefits.[40] Nevertheless, the Tikorangi action group has continued to meet, plan, coordinate protest actions and make submissions at council hearings on resource consents for new wells, tightening regulations on production operations, and minimizing infrastructure "improvements" that are rapidly industrializing their rural landscape.

Confronting Oil and Gas Corporations

Critics of globalization maintain that the increased power of financial institutions and transnational corporations through globalization has undermined the ability of governments to hold corporations accountable or control the trajectory of national development. This in turn has a detrimental effect on how much influence local communities retain over on their own social and economic development. [41] These days governments are just as likely to act to weaken or restrict local democracy and dictate the parameters of regional development in the "national [corporate] interest" as has been the case in New Zealand.[42]

There appears to be growing recognition around issues like oil and gas development, climate change and transitioning to a low-carbon economy that grassroots democracy needs to be reinvented if these issues are to be addressed effectively. These are not philosophical debating points. Kelsey[43] observes that the global domination of powerful transnationals is being challenged by forces at national and international levels determined to protect the rights of workers, local economies and the environment. "The rapid spread of new technologies that manipulate nature for profit [e.g., fracking, coal seam gas, tar sands extraction] has galvanized indigenous peoples, environmentalists and consumers" into action.

Rural communities, urban neighborhoods, and indigenous peoples threatened by mining and oil and gas extraction at some stage reach a crossroads. Either the ideological views and material interests of opposing factions prove so irreconcilable that they override long-standing community cohesion, and factions end up dealing with extractive corporations in their own ways (i.e., the corporate divide-and-conquer tactic). Or leaders and factions recognize that to protect their shared community identity and control their own development they have to reach some kind of consensus and confront corporations and their government supporters from a position of collective strength. Boomtown case studies indicate that the former path can lead to detrimental social and economic changes, lasting divisions within the community, and an exodus of disaffected citizens. The latter path stands a better chance of the whole community "winning" by, for example, passing a referendum blocking a project or successfully negotiating a more beneficial arrangement for the community. But it can equally result in the whole community losing if a community referendum, protests, or other initiatives are thwarted by government intervention or corporate legal action.

Some communities and First Nations have chosen to pursue partnership arrangements with petroleum companies. The hope is that the formal relationship will turn out to be equitable in practice and recognize the community or first nation's right to self-determined development.[44] The risk is that such partnerships may not lead to promised economic and social well-being, but instead result in internal divisions that end up playing into the hands of corporate or government efforts to co-opt the community or first nation.[45] Gedicks documented how large mining companies in ten different countries adopted surprisingly similar divide-and-rule tactics to subvert indigenous opposition and gain access to their land for their extractive projects.[46] Other communities and indigenous tribes have chosen to resist rather than seek accommodation with extractive companies because they believe companies ultimately have their own corporate interests at heart and their operations are likely to do more harm than good. Confrontation can range from publicity campaigns, to answering company and government PR propaganda, to exposing company tactics and cover-ups, to undertaking legal and protest action.[47]

Why are such localized protests and confrontations significant? McAdam and Boudet argue that when one examines most national struggles they turn out to be "aggregations of local movements" or networks of community protest actions.[48] A local spark can ignite a much wider movement for change. In Australia, the Lock the Gate Movement arose out of protests by Queensland small holders over a coal seam gas project in their district. It grew rapidly into a national movement when it became clear there was need to coordinate and communicate among a growing number of separate rural landowner protests. Similarly Indigenous Australian communities have organized themselves nationally to share knowledge, obtain information, gain assistance from academic institutions, and confront big mining corporations over new mining developments.[49] Their position has been strengthened by passage of the Native Title Act 1993 which requires exploration companies to settle heritage protection, land access, and other traditional land owner issues before an exploration permit is awarded.[50]

In New Zealand the spontaneous establishment of a number of oil-free and frack-free community action groups when international companies began taking up government exploration permits catalyzed an informal national network in opposition to the petroleum industry. These groups communicate regularly, have linked up with established environmental organizations, and occasionally undertake coordinated actions such as blockading government or oil company offices, protesting at petroleum conferences or council meetings, and organizing petitions such as Greenpeace's "Block the Block Offer 2016" campaign.

On the one hand, as we have noted, the National government has been keen to seek collaborative relations with mainstream environmental and recreational groups like Forest and Bird, Fish and Game, and EDS around key environmental issues. In October 2015 for instance EDS hosted the Australia-New Zealand Climate Change and Business Conference. There was broad consensus among community, environmental, local government, and business representatives about the need for a permanent "Climate Forum" to bring together all the key sectors at a national level to chart a course to transition to a low-carbon future. The forum idea was "welcomed" by Climate Change Minister Tim Groser and Energy Minister Simon Bridges[51] but no further action was taken. Ten months later newly appointed Climate Change Minister Paula Bennett announced the government would set up its *own* task force to look at the issues. Major national protest organizations like Greenpeace and groups within the ECO coalition viewed this as symptomatic of the National government's reluctance to revisit the objectives of its Business Growth Agenda since it would threaten oil and gas development and impose increased costs in having to deal with climate change. Apparently sensing growing public backlash over the government's obstinacy, these "extremist" elements called for an escalation in protest around events such as the annual Advantage NZ Petroleum Summit to outright civil disobedience on a scale similar to the anti-apartheid protests against the 1980 Springbok Rugby tour.[52]

Meanwhile, in Taranaki a new grassroots organization called Taranaki Energy Watch (TEW) had emerged to confront the established oil and gas industry, working in collaboration with groups like Climate Justice Taranaki. The group, headed by ex-farmer and high-profile activist Sarah Roberts, focused on "sharing information and supporting local communities to protect their health and the environment from the effects of oil and gas exploration and production in Taranaki and New Zealand."[53] TEW helped landowners stand up to oil companies seeking access to their land, organized protests outside production facilities, and made submissions on consent hearings and district plan changes favoring the oil industry, as well as on RMA amendments.

Indigenous Self-Determined Development and Resistance

Oil companies and the National government have made much of the potential benefits of petroleum development for Māori, citing the Taranaki region as a case in point. There are approximately eleven iwi in the region, three of whom have signed formal relationship agreements such as Ngāti Ruanui's agreement with TAG Oil. Because of local geology, only some iwi primarily on the east side of revered Mt Taranaki have been directly affected by oil and gas developments. The industry dates back to the late nineteenth century and in some districts Māori have been employed as oil workers, truck drivers, and caterers for several generations. Some have become highly skilled rig workers and travelled the world. As a Māori environmental activist observed,[54] people have come to accept the industry as a fact of life. They may not like what it's doing to their land or way of life, but they have to look after their families and put food on the table.

Some groups have been reluctant to enter into working relationships with exploration and production companies. Others have tried to stop the intensification of the industry in their rohe (district). But even though they have made submissions about new drilling sites under the RMA, they realize there is little else they can do under current government policies and regional planning regulations. Two iwi have managed to use Waitangi Treaty settlement negotiations with the Crown to grandfather in clauses requiring companies to consult them on any land or environment effects of new projects.

While there has been little iwi-level opposition like Ngāti Porou's rejection of petroleum development on the East Coast, some hapū have engaged in resistance action to try to block exploration and drilling. Otaraua Hapū has been active around Tikorangi in North Taranaki, led by outspoken chairman Rawiri Doorbar. Whenever there is a drilling proposal in the area, hapū representatives attend meetings, hold face-to-face discussions and present submissions at consent hearings. In 2009 Otaraua gained notoriety for occupying a proposed Greymouth Petroleum well site for over three months. At the end of 2015, Heritage New Zealand and Otaraua hapū appeared in court objecting to Greymouth's plans to drill near the grave site of paramount chief Wiremu Kingi Te Rangitaake.[55] As a member of the hapū explained, "That's one of the companies the people hate the most. They call them cowboys. They have just a really bad relationship with the community."[56] Based on their experiences, Otaraua chairman Doorbar advised landowners attending the January 2015 ECO conference to "stop them at the gate!"

Another group around the South Taranaki settlement of Patea has been similarly active in trying to block petroleum development in recent years. They have adopted the standard practice when a new well site proposal is notified of lodging a submission with the local council, speaking out at public meetings, and organizing protest actions. Toward the end of 2015 several members were arrested for pulling up the survey pegs of a major oil company. When the matter went to court in March 2016, a group of about

fifty supporters—several waving Māori sovereignty flags—gathered outside
and violent scuffles broke out with police.[57]

In Northland, Norwegian oil company Statoil experienced ongoing difficulties with Māori activists over plans for offshore exploration. When it
was first mooted that the National government was interested in expanding
extractive activities as part of its economic growth agenda, iwi and hapū leaders in Northland decided to embark on a proper consultation on mining with
their people. The process took almost two years, but eventually the collective
consensus was against mining while leaving the final decision in each hapū's
hands. When the government organized several hui to "consult" on an offshore block offer, the message back to the Minister according to iwi leader
Dr Margaret Mutu was "Absolutely no way are you to come up into our area
and mine." However, Mutu says,

> They just ignored us. But we're quite used to this sort of treatment. It doesn't
> surprise us. So what they did—a typical maneuver of governments in this
> country: *divide-and-rule!* Here there was only one iwi chairperson they could
> get to drive this thing through. So you've got five iwi and maybe about thirty
> hapū involved, and you get one chairperson to try and override everybody. And
> as always with divide-and-rule, the result was major disputes. . . . But eventu-
> ally, they decided to open up Te Reinga Basin for a block offer, and we were
> horrified. Because we thought of all the areas of the country you wouldn't go
> into, this is Te Ara Wairua—the pathway of the spirits, where the spirits from
> anywhere around the country follow a pathway that is guarded carefully by us,
> that is our job. And when we questioned what benefits were going to come to
> iwi, there were none. But there were substantial risks.[58]

According to Greenpeace Norway spokesperson Martin Norman, Statoil
goes in trying to be "friendly Norwegians wanting to do good but not understanding the societies they're going into. . . . It seems to have happened
in Canada and seems to be happening in New Zealand."[59] In August 2015
Statoil was accused of organizing secret meetings with selected iwi and hapū
leaders. Māori activist Mike Smith explained, "You don't go consulting with
iwi by going around picking off individuals, meeting with them on the quiet,
that's not how we do it! . . . They are desperate to get a social license signed
off from the people up north."[60] Smith headed a delegation earlier in May
2015 to Norway to meet with the Sami indigenous people's parliament and
address Statoil's annual meeting. The delegation warned the Statoil's AGM
the company did not have permission from Māori to explore in their tribal
waters. Three of five Northland tribal groups officially came out against
Statoil's plans. Company executives were aware of the growing opposition[61]
and apparently decided to take pre-emptive action by arranging clandestine

meetings with iwi leaders. Statoil's Pal Haremo justified the move, pointing out that he had already visited New Zealand six times: "We're concentrating on five major iwi and then leave it to chairmen and iwi leaders to engage with others. We just can't have hundreds of meetings [and protests] each time we have new information."[62] Evidently Haremo expected iwi leaders to convey information to "their people" on the company's behalf and quell the protests.

RESISTING PETROLEUM DEVELOPMENT
ON THE EAST COAST

As yet there has not been the extensive oil and gas exploration and development on the East Coast that would spark sustained community and tangata whenua resistence, as has happened in Taranaki. Though government legislative reforms and industry promotional activities continue, the temporary downturn in global oil markets have meant only a few exploratory wells have been drilled on the East Coast. That has resulted in the existence of a loose network of individual activists and anti-oil groups who communicate and take action sporadically as the occasion warrants. When a new development is proposed (e.g., a well site application), a minister visits, or there is an opportunity to make a submission to Council (changes to a district or regional plan relevant to petroleum development) the network is sparked into action. For example, Napier-based Don't Frack the Bay and Frack Free Tararua have organized fracking information meetings and protests against ministerial visits, water issues, and establishment of TAG Oil's field office in Napier and TAG's exploratory well near Dannevirke.

Frack Free Tairāwhiti[63] based in Gisborne operated in a similar ad-hoc fashion, communicating and organizing events through a loose network of concerned citizens, academics, Māori activists, and green-leaning district councilors. TAG/Apache's presentation to the Gisborne District Council in early 2012 and their Punawai-1 resource consent application were catalysts to the establishment of this network. When the consent application was lodged, there was apprehension among residents of the small settlement of Whatatutu near the proposed well site over the potential impacts. Councilor Manu Caddie, spokesperson for the Council's green faction, helped organize a public information hui at the local marae which attracted not only locals but a contingent of concerned people from throughout the Gisborne district. The hui was addressed by visiting Professor Avner Vengosh of Duke University who shared US research on some of the negative effects of unconventional drilling and fracking. Councilor Manu Caddie encouraged participants to "learn the facts about exploration" and "let your community know your concerns" through the media, public meetings, and the Internet. Within weeks, a group calling itself Frack-free Tairāwhiti (FFT) was exchanging emails and planning meetings, symposia, and other actions. Councilor Caddie issued media releases and other members began writing letters to the editor about the dangers of unconventional oil and gas development.

A further informational meeting was arranged in the township near the well site six months later in June 2012. Council Environment and Policy Manager Hans van Kregten was invited to attend and explain the resource consent process. He refused to be drawn on whether TAG/Apache's resource consent application for Punawai-1 would be publically notified so the public could make submissions. It was after that hui that FFT circulated a petition calling for the consent process to be notified which received close to 2,000 signatures. To raise awareness about the petition and the issue of fracking, FFT organized a protest march of several hundred people down the main street of Gisborne. A month later, the group arranged a public seminar in the District Council chambers featuring Drew Hutton, coordinator of the Australia's Lock the Gate movement who had been touring New Zealand at the invitation of anti-fracking groups.[64] Coinciding with the seminar, FFT held a series of screenings of the controversial documentary *Gasland.* Following Hutton's tour, several anti-fracking groups joined the Australian Lock the Gate network and a member of Hawke's Bay's Don't Frack the Bay became the movement's New Zealand coordinator.

Protest action took a creative turn later in 2012 when government awarded TAG/Apache another exploration permit near Gisborne which included a portion of the rich horticultural and wine growing land around the city. Neither the District Council, growers, nor the general public were clear where the permit boundaries actually were. A few weeks after the permit was announced, official-looking road signs suddenly appeared around Gisborne informing drivers they were entering or leaving "Apache Territory." Anti-fracking activists had accessed technical data about the permit on NZPaM's website, used GPS to locate the exact boundaries, replicated Apache's company logo on the signs, and erected them in the dark of night. When news of the signs hit the media, tangata whenua were outraged. Despite the furor the signs remained in place for more than a fortnight. District Council officials eventually concluded they hadn't been erected by the Council, the JV, or central government and had them removed. Coincidently, at the time the signs appeared Apache Corporation was dealing with violent protests by Argentina's Mapuche people who had been fighting an ongoing battle with the company over access to tribal lands.[65] A few months later, Apache withdrew from the JV and left the East Coast, possibly to avoid further reputational damage from yet another high profile confrontation with indigenous people.

FFT met with the Commissioner for the Environment when she visited Gisborne later in 2012, and continued to organize symposia and raise concerns around unconventional oil and gas development through a new website, letters to the editor and various public appearances by Councilor Manu Caddie who became the group's de facto spokesperson.[66] Caddie was frequently interviewed by the national media, and in November 2013 appeared in a televised debate over oil and gas development on TV3's VOTE program. It was the same week that Northland Māori threw NZPaM officials off their marae. True to form, Energy and Resources Minister Simon Bridges responded a few days later with an op-ed piece on the safety of the industry and the need for "environmentally responsible development."[67] The *Gisborne Herald* followed with an editorial a fortnight later arguing "fossil fuels will be with us for a long time to come."

FFT made presentations to the Gisborne District Council and the Council's Environment and Regulations Committee when TAG Oil applied for a second consent for at the Punawai-1 site in early 2013 and again in 2014 for their Waitangi Valley-1 well. Once again they called for all extractive consents to be publically notified. This time, Councilor Caddie raised the issue directly with Council CEO Judy Campbell, pointing out that the Parliamentary Commissioner for the Environment had recommended in her final 2014 fracking report that councils exercise their discretion to involve the public in controversial petroleum consents. The majority of councilors chose to ignore the PCE's recommendation, voting to continue the Council's policy of not publically notifying extractive consents if council staff judged the effects identified in a drilling company's environmental impact statement were "less than minor."

In spite of their disadvantaged position vis-à-vis central government policies and local government gatekeeping, citizen action groups occasionally achieve unexpected success. On November 20, 2015, the Gisborne District Council voted to oppose petroleum exploration in the offshore Pegasus and East Coast basins as proposed in the government's latest Block Offer 2016.[68] The vote, which came as a bombshell to many in the local business community, the editor of the Gisborne Herald and no doubt some National government ministers, endorsed a decision by the Council's Environmental Planning and Regulations Committee the previous month. The committee had voted to support a petition from over six hundred citizens organized by a network of residents opposed to offshore drilling including FFT activists. The petition was presented to the committee by a local anthropology student, a water conservation specialist, and an iwi environmental advocate. It was obvious from their presentations that all had done their homework, citing evidence and examples of the risks around unconventional and deep-sea drilling. Similarly during the regular public forum section at a full Council meeting the following month, a presentation was delivered by a Māori environment, and resource management tutor from Te Waananga o Aotearoa.[69] It was a classic example of citizen science in action and no doubt influenced the outcome of the Council's vote.

Councilor Rehetta Stoltz captured the mood of more moderate councilors who voted against deep-sea drilling in the two offshore basins near Gisborne. She said she usually leaned toward responsible economic development, but in this case there was not enough evidence that the economy would benefit from petroleum exploration and drilling and there was a chance of some big environmental problems. Unfortunately, the Council's submission would not deter the government from going ahead with the block offer:

> I am frustrated that our community is tricked into thinking that we can influence our government at this stage. There is such a strict legal frame where we are obliged to issue consents for exploration that the Government sells to the highest bidder. We have no power really.[70]

Tairāwhiti anti-fracking activists and citizen petitioners, like those in other communities around the country who had lodged similar coordinated petitions with the assistance of Greenpeace, would no doubt have begged to differ with the councilor. In the end two regional councils, three district councils and eight Māori iwi authorities on the East Coast lodged objections or requests for special conditions to the latest block offer.[71] The government answered these objections and proceeded to award exploration permits anyway, but the genie was out of the bottle. A national survey by PEPANZ found that East Coast residents were the most negative about oil and gas development of any region in the country.

CONCLUSION

Patricia Widener[72] concluded from her comparative research on community responses to oil and gas development that most such developments are imposed on communities without giving them the opportunity saying no. This is certainly the case with respect to current New Zealand government policy settings, legislative reforms, and the annual block offer process with its perfunctory consultation process. As the residents of Tikorangi township learned, when oil and gas operations expand rapidly there may be little time to respond. They can be caught off guard with insufficient information, or reliant on municipal authorities who may initially favor such developments.

The situation is exacerbated when the objectives of the project and government policies supporting it are at odds with community and indigenous values, aspirations and/or environmental standards.

Communities are changed when a "boom" occurs, for good or ill depending on the perspectives and material interests of local residents. Communities and indigenous nations do not respond uniformly to extractive industry developments because (a) people have different views, values, and interests that ultimately have to be negotiated; and (b) extractive companies frequently apply divide-and-rule tactics to try to build majority support for a proposed development.

For communities attempting to reach a consensus on a way forward rather than become fragmented by corporate and/or political agendas, or for community action groups who are trying to lead the way in resisting petroleum extraction, this chapter has identified a number of activities which those in similar circumstances have utilized. For New Zealand the common themes running though these activities have to do with the power of independent evidence, citizen-driven public scrutiny, and maintaining strong communication networks with allies outside the community.

Organized social action against oil and gas development involves more than public protest. In most New Zealand communities only a handful of committed activists have engaged in such actions. Concerned citizens may become involved and join the debate as deep-sea or unconventional operations suddenly threaten a community or sensitive environmental area as they have in parts of Taranaki. Even in these circumstances, before protest and civil disobedience erupts there is typically a gradual build-up of information gathering, meetings and public education, counter-propaganda campaigns, lobbying, organizing, and alliance building. New Zealand has a long-established tradition of open governance and government/civil sector engagement. Environmental organizations, Māori activists, and local anti-fossil fuel groups have concentrated their efforts at a national level on influencing government policy and legislation through concerted lobbying, submissions on pending legislation (especially the RMA) and policy discussion documents. At a local level, anti-fossil fuel groups have tended to focus on making submissions on Long-Term Council Plans and regional resource management plans, and lobbying and criticizing municipal council policies and decisions (e.g., Gisborne Council's failure to open resource consent processes to the public). As Taranaki and East Coast environmental groups and Māori activists have discovered, such efforts may achieve short-term results but often the struggle to block or substantially modify development projects is unsuccessful.

Institutionally, this has resulted in a protracted and increasingly acrimonious wrestling match as the state backed by Big Oil undertakes ever more authoritarian manoeuvers to push through legislative changes, override or influence local government development planning and environmental consenting processes, minimize "inefficient" citizen engagement, and placate indigenous groups. In turn this has catalyzed more well-organized, committed, and forceful responses not only from established environmental organizations but growing numbers of concerned citizens who have till now remained at the margins of the struggle. As the petroleum industry recovers from the global oil market slump and gears up for a renewed round of offshore and onshore unconventional exploration, there is potential for such encounters to become increasingly volatile as issues around expansion of the petroleum industry clash with demands to preserve "our clean, green environment" and urgently transition to a low-carbon economy.

NOTES

1. Korten, *When Corporations Rule the World,* 124. John Mitchell, "A New Political Economy of Oil," ISSN 1062-9769, *Quarterly Review of Economics and Finance*, Volume 42, 2 (2002): 251–272.

2. Bebbington, "Extractive Industries, Socio-Environmental Conflicts and Political Economic Transformations in Andean America," 16.

3. Doug McAdam and Hilary Boudet, *Putting Social Movements in their Place Explaining Opposition to Energy Projects in the United States, 2000–2005* (Cambridge: Cambridge University Press, 2012).

4. E.g., Patricia Widener, "Community Responses to the Social, Economic and Environmental Impacts of Oil Disasters and Natural Resource Extraction" (presentation to a public seminar on issues around oil and gas development, Gisborne District Council chambers, Gisborne, New Zealand, October 16, 2013); Bebbington, "Extractive Industries"; Brasier et al, "Residents' Perceptions"; Richard Krannich, "Social Change in Natural Resource-Based Rural Communities: The Evolution of Sociological Research and Knowledge as Influenced by William R. Freudenberg," *Journal of Environmental Studies and Sciences* (2011), doi: 10.1007/s13412-011-0051-y.

5. Widener, "Community Responses."

6. See for example Andrew A. Rosenberg, Pallavi Phartiyal, Gretchen Goldman and Lewis M. Branscomb, "Exposing Fracking to Sunlight: the Public Needs Access to Reliable Information about the Effects of Unconventional Oil and Gas Development in Order for it to Trust that Local Communities' Concerns Won't Be Ignored in Favor of National and Global Interests," *Issues in Science and Technology* 31, 1 (Fall 2014), http://issues.org/31-1/exposing-fracking-to-sunlight/.

7. Wilber, *Under the Surface,* 34.

8. Patricia Widener, "A Protracted Age of Oil," 843.

9. Widener, "A Protracted Age of Oil," 843.

10. Gretchen Goldman, "Colorado Towns Pass Fracking Moratoria, Bans Despite Big Spending by the Oil and Gas Industry," *The Equation*, Union of Concerned

Scientists, November 6, 2013, accessed November 12, 2014, http://blog.ucsusa.org/gretchen-goldman/colorado-towns-pass-fracking-moratoria-bans-despite-big-spending-by-the-oil-and-gas-industry-293.

11. Thomas Kinnaman, "The Economic Impact of Shale Gas Extraction: A Review of Existing Studies," *Ecological Economics* 70 (2011): 1243–1249, accessed September 4, 2013, from Bucknell University Digital Commons website, http://digitalcommons.bucknell.edu/fac_pubs/5.

12. "Study Puts Price Tag on Dairy," *Radio New Zealand*, December 24, 2014.

13. Terrence Loomis, "Indigenous Populations and Sustainable Development: Building on Indigenous Approaches to Holistic, Self-determined Development," *World Development* 28, 5 (2000): 893–910. Terrence Loomis, "Māori Initiatives in Sustainable Development," *The Journal of Aboriginal Economic Development* 3, 2 (2003): 81–88. Terrence Loomis with John Mahima, "Māori Community-based Sustainable Development: A Research Progress Report," *Journal of Development Practice* 13, 4 (2003): 399–409.

14. Personal communication with Professor Avner Vengosh March 15, 2012.

15. Under the Local Government Act 2002, councils and communities were responsible for developing long-term plans to achieve balanced social, economic, environmental, and cultural outcomes. Such a requirement led some councils to experiment with "holistic" cost/benefit analysis.

16. Widener, "A Protracted Age," 9.

17. Discussion with course tutor Tina Ngata, October 2014.

18. Drew Cadenhead, Transcript of TAG Oil's FQ4-2013 conference call, April 2014, 8, accessed September 20, 2014, www.tagoil.com. "Spindletop" refers to a Texas well that erupted with oil in 1901, touching off an oil boom.

19. Widener, "A Protracted Age," 10.

20. E.g., Wilber, *Under the Surface*, 44–45. See also McGraw, *The End of Country*.

21. Elizabeth Ridlington and John Rumpler, *Fracking by the Numbers: Key Impacts of Dirty Drilling at the State and National Level*, Environment America Research and Policy Center, Boston, 2013, accessed November 9, 2013, http://www.environmentamerica.org/sites/environment/files/reports/EA_FrackingNumbers_scrn.pdf. See also Catskill Mountainkeeper, http://www.catskillmountainkeeper.org/.

22. Christopher Bateman, "A Colossal Fracking Mess," *Vanity Fair*, June 2010, http://www.vanityfair.com/news/2010/06/fracking-in-pennsylvania-201006

23. Keely Kidner, "Beyond Greenwash," 2; 35.

24. Catskill Mountkeeper, the Natural Resources Defense Council, Earthjustice, Environment America, and Cleanwater Action are examples in the United States. Climate Justice Taranaki (https://climatejusticetaranaki.wordpress.com) and Oil Free Otago (www.oilfreeotago.com) are New Zealand examples.

25. "Rallies around Country against 'Corporate Takeover,'" *NewstalkZB*, September 1, 2013, accessed September 19, 2014.

26. http://www.generationzero.org/.

27. http://wiseresponse.org.nz/.

28. *Get Free* was launched by Greenpeace in September 2013, with several high-profile New Zealanders fronting the online campaign.

29. See also Diprose, Thomas and Bond (2016:166 ff) for a discussion of recent resistance activities by New Zealand environmental movements and anti-oil community groups.

30. Kate Stringer, "Two Towns Battle Colorado for Freedom to Ban Fracking," *Yes! Magazine*, July 28, 2016, accessed August 28, 2016.

31. See Javier Arellano-Yanguas, "Mining and Conflict in Peru: Sowing the Minerals, Reaping a Hail of Stones," in *Social Conflict, Economic Development and Extractive Industry: Evidence from Latin America*, ed. Anthony Bebbington (London: Routledge, 2012), 89–111.

32. http://www.eco.org.nz/

33. "Warts and All Exposé of Taranaki's Oil and Gas," *Scoop News*, January 27, 2015, accessed January 28, 2015. "Taranaki's Dirty Secret Is Out!" *Scoop News*, February 2, 2015, accessed May 3, 2015.

34. "Community Joint Statement – A Call to Ban Fracking in New Zealand," ECO, published February 1, 2015, https://climatejusticetaranaki.wordpress.com.

35. Union of Concerned Citizens, *Science, Democracy and Fracking: A Guide for Community Resident and Policy Makers Facing Decisions over Hydraulic Fracking*, Cambridge, MA, 2013, http://www.ucsusa.org/sites/default/files/legacy/assets/documents/center-for-science-and-democracy/fracking-informational-toolkit.pdf.

36. See Widener, *Oil Injustice.*

37. Confidential interview with a spokesperson for Climate Justice Taranaki on December 5, 2015. The interviewee was advised the discussion was being recorded and consented to the information provided being available for possible publication on condition of anonymity.

38. E.g., see Tikorangi Newsletter on http://jury.co.nz/petrochem.

39. "Big Oil Spooks Tiny Town," *New Zealand Herald,* March 3, 2013, accessed October 30, 2015.

40. "Big Oil Spooks."

41. E.g., Karliner, *The Corporate Planet*, 217. For other examples see *Transnational Corporations* 18, 1 (April 2009), United Nations, Geneva.

42. Jane Kelsey, *Reclaiming the Future.*

43. Jane Kelsey, *Reclaiming the Future,* vii.

44. For recent South American case studies of local protest movements against mining companies, see Bebbington, *Social Conflict.*

45. For Asian and Pacific examples, see Glenn Banks, "Activities of TNCs in Extractive Industries in Asia and the Pacific: Implications for Development," in *Transnational Corporations* 18, 1 (April 2009), United Nations, Geneva: 51–68. Also Bethany Haalboom, "Confronting Risk: A Case Study of Aboriginal Peoples' Participation in Environmental Governance of Uranium Mining, Saskatchewan," *The Canadian Geographer / Le Géographe Canadien* 4, 58 (3) (2014): 276–290, DOI: 10.1111/cag.12086.

46. Al Gedricks, *Resource Rebels: Native Challenges to Mining and Oil Companies* (Cambridge, MA: South End Press, 2001).

47. See Katia Opalka, "Oil and Gas: the View from Canada," *Natural Resources & Environment* 26, 2 (Fall 2011): 37.

48. McAdam and Boudet, *Putting Social Movements in Their Place*, 69.

49. See Jon Altman and David Martin, eds., *Power, Culture, Economy;* also Benefict Scambary, *My Country, Mine Country: Indigenous People, Mining and Development Contestation in Remote Australia*, CAEPR Research Monograph 33 (Canberra: ANU Press, 2013).

50. "Mosman Oil and Gas to Withdraw From Officer Basin Project after Failing to Meet Native Title Act Requirements," *OilVoice*, January 18, 2016.

51. *Environmental Defence Society Latest News*, December 2015, http://www .eds.org.nz/

52. Email from Steve Abel, "Time for Peaceful Civil Disobedience," Greenpeace, February 23, 2016, nyoung@greenpeace.org.nz.

53. http://taranakienergywatch.org.nz/about-us/.

54. Confidential interview with a local Māori environmental activist February 9, 2016. The interviewee was advised the discussion was being recorded and consented to the information provided being available for possible publication on condition of anonymity.

55. "Oil and Gas Permits a Massive Hypocrisy," Climate Justice Taranaki newsletter, December 16, 2015, https://climatejusticetaranaki.wordpress.com/2015/12/16/oil-and-gas-permits-a-massive-hypocrisy/.

56. Māori environmental activist, February 9, 2016.

57. "Hawera District Court Closed After Protest," *Taranaki Daily News*, March 1, 2016, accessed March 5, 2016.

58. Dr. Margaret Mutu, "Ngāti Kahu Kaitiakitanga meets Statoil and Shanghai CRED" (presentation to the Symposium on Māori Engagement with Extractive Industry, Faculty of Law, Auckland University, June 12, 2015).

59. "Norwegian Oil Hunters—Coming to NZ," *New Zealand Herald,* October 16, 2015.

60. "Norwegian Oil Explorer Accused of Divide-And-Rule Tactics in Northland," *Radio New Zealand*, August 27, 2015.

61. "Northland Iwi Face Statoil over Oil Drilling Plans," *New Zealand Herald*, August 16, 2015, accessed August 18, 2015.

62. "Norwegian Oil Hunters."

63. Te Tairāwhiti," the traditional Māori name for the region around Gisborne.

64. See www.lockthegate.org.au/.

65. "Mapuché Woman Acquitted by Intercultural Jury," *The Argentina Independent*, November 5, 2015, accessed November 29, 2015. One of the protest leaders Relmu Namku stated during a court case, "It pains me to be sitting here while the real culprits [Apache Corporation] are not, because they have power, they have money, and there is total impunity for them."

66. For example, Cr Caddie delivered a presentation against oil and gas development at the October 2013 symposium organized by Hastings Mayor Lawrence Yule. Personal communication with Cr Caddie, October 29, 2013.

67. Simon Bridges, "Committed to Upholding Oil and Gas Safety Record," *Gisborne Herald*, November 9, 2013, accessed August 21, 2015.

68. "Councillors Vote 8-5 Against Offshore Oil Drilling," *Gisborne Herald*, November 20, 2015, accessed November 21, 2015.

69. The address can be viewed on http://www.authorstream.com/Presentation/tina282217-2660219-three-myths/.

70. "Councillors Vote 8-5." TAG Oil's ex-chief executive Garth Johnson and ex-chief operating officer Drew Cadenhead obviously expected the Government to proceed with the Block Offer regardless of the Council's vote. They established a new venture called Hydrate Resources Corporation to exploit the gas hydrates in the two East Coast basins.

71. Christine Walsh, "Crown Going Ahead with Offering Offshore Blocks Despite East Coast Opposition," *Gisborne Herald*, March 26, 2016, accessed March 27, 2016.

72. Widener, "Community Responses."

Conclusion

Corporatist democracies like New Zealand, who are seeking to boost economic growth through intensified resource extraction at a time of increasing public concern over climate change and the harm caused by unconventional oil and gas development, are a useful site for examining the structural (institutional) effects of the struggle between state/corporate partnerships promoting expanded petroleum development and environmental NGOs and communities opposed to such development. Bebbington, Kirsch, and Benson and Kirsch[1] identified several gaps in the literature on social conflict, economic development, and the extractive industry that need to be addressed. In the first Chapter I indicated this study would focus on three particular issues:

1. The strategies transnational corporations use to defend themselves and avoid accountability for the externalized costs of their operations, as well as how they influence state policies and state responses to protest action,
2. The maneuvers states adopt to promote extractive development and deal with critics and protests, and
3. How protest action by communities and environmental groups, and government responses to these actions, affects institutional relations between extractive corporations, states, and the community sector.

In order to account for the effects of community and indigenous environmental activism on state policy-making and government/petroleum industry relations in New Zealand (my third research issue), this study adopted a Regulation Theory framework to examine key changes in institutional relations and values.

STRATEGIES OF PETROLEUM CORPORATIONS TO PROMOTE AND DEFEND THEMSELVES

The enormous economic power petroleum and mining companies have acquired through globalization has enabled them to dominate most aspects of political life and public opinion in resource rich countries. In some third-world countries this had created a phenomenon called the "resource curse" where extractive industries effectively control both the state and the economy. In countries with more transparent political institutions and mixed economies such as the United States, Canada, the United Kingdom, Australia, and New Zealand the literature suggests oil and gas corporations have adopted remarkably similar strategies to promote the industry, avoid punitive regulatory regimes that would force them to pay for the harm they do, and defend themselves against criticism and protest. These strategies are often associated with typical tactics which corporations and their trade associations employ depending on circumstances (table 7.1).

Table 7.1. Oil and Gas Industry Strategies and Tactics.

Strategies	Tactics
Influencing and collaborating with governments	Patronage and lobbying Helping write user-friendly legislation and regulations Revolving door arrangements Externalizing costs
Selling the industry to the "silent majority"	Exaggerating the sector's reserves and future energy role Greenwash: proclaiming corporate stewardship Showcasing corporate social responsibility Acknowledging problems, covering up harm
Controlling affected councils, communities, and indigenous groups	Trust-building consultations, speaking engagements, etc. Utilizing collaborative "stakeholder partnerships" Buying community support
Defending company and industry practices	Manufacturing debate, controlling public discourse "Harnessing" mainstream media Utilising disinformation campaigns and "experts" Funding foundations, think tanks, and advocacy groups "Harnessing" social media Infiltrating the education system
Combatting opponents	Countering protest movements Co-opting environmentalists, marginalising extremists Direct force, court action, law enforcement

Operationally, the New Zealand petroleum industry (local companies, overseas-based juniors, and a few major international corporations) is comprised primarily of a few fields off the coast of the Taranaki region and an onshore unconventional sector that has emerged in the past decade. The election of the Fifth National government under John Key resulted in a shift in government policy toward actively marketing the country's prospectivity on the international stage to compete with other "frontier" states, establishing a legislative and regulatory regime to facilitate oil and gas expansion, and promoting the industry to the New Zealand public. With a relatively small economy, a modest-sized petroleum industry, and a political arena occupied by a diversity of competing interest groups, it is perhaps not surprising that promotion and defense of the oil and gas industry has taken the form of a loose network of organizations and individuals linked together by a broadly shared ideology rather than a large, well-endowed "machine" as in some larger petroleum-producing countries.

The government's agenda for expanded resource extraction outside the Taranaki region has served as a catalyst to public debate, political confrontation, and social conflict which has placed oil and gas companies and their trade association PEPANZ in a conundrum. Until recently, they have been able to carry on business as usual with minimal public controversy, relying on low-key strategies aimed at maintaining the goodwill of the "silent majority"[2] and influencing state policies. The National government's intensified industry promotional campaign has increased opportunities (and pressures) for greater collaboration between the industry and the state. It has added impetus to organized environmental opposition and sparked local anti-oil/anti-fracking protests. Oil companies and PEPANZ initially responded to this opposition by attempting to engage and confront environmental critics through manufactured debate, "managing" protest, maintaining PR spin about the industry's benefits and safety, and going through the motions of "consulting" with communities and Māori.

However, the industry's tactics and rhetoric served to energize and validate a growing environmentalist and anti-fossil fuel movement particularly during the Commissioner for the Environment's investigation into fracking and the lead-up to the 2014 national election. Influenced by advice and experience from overseas (such as the de-escalation tactics of Tisha Schuller's Colorado Oil and Gas Association) as well as the government's maneuvers to engage with *moderate* environmentalists around resource management issues, the petroleum industry shifted its approach from confrontration to acknowledging public concerns about the environment and climate change, and dialoguing with moderate environmentalists and outdoor recreational organizations while for the most part ignoring extremist protesters. In terms of Benson and Kirsch's three phases of harm industry defense,[3] the industry and PEPANZ in particular shifted the core of its approach from phase 1 strategies (business as usual, lack of engagement around externalities, and problem-denial) to phase 2 (problem acknowledgment, token engagement, and accommodation).

That is, a shift from reliance primarily on "factual" information dissemination, "reasonable" public debate, and claims the industry was complying with government safety and environmental regulations to strategies aimed at demonstrating greater transparency, apparent accountability, and managed engagement with critics. The global oil market slump and a reduction in unconventional and offshore exploration outside of Taranaki have moderated impacts on local communities and Māori iwi. But they have also given the industry an opportunity to operate under the radar through covert strategies (e.g., funding academic centers and GNS research, industry education penetration, "culturally appropriate" consultation, and arranging development MOUs with iwi). However, there has been growing support for calls by environmentalists and local anti-fracking/fossil fuel activists for an urgent transition plan to a clean-energy economy from scientists, LGNZ, and sustainable business organizations, bringing pressure on the industry to at least contemplate a proactive move to Benson and Kirsch's phase 3 (strategic engagement and crisis management) in order to avoid the threat of catastrophic company loss, industry collapse, or complete loss of legitimacy.

At the same time, petroleum companies operating in New Zealand have sought to influence public policy through a variety of formal channels including producing briefing papers, commenting on ministerial discussion papers, and making submissions to select committees. On rare occasions an industry representative or consultant has been asked to sit on a ministerial advisory group. Such tactics are easier to employ in a capital city which tends to mimic a small town, even though there are 121 MPs and a large public service. The political community is woven together by interlocking networks of work contacts and "mates"—bureaucrats, industry lobbyists and consultants, corporate staff, ministers, and executives—who maintain contact via phone, email, socializing, coffee meetings, and lunch chats. Petroleum industry representatives have been able to influence policy development more effectively through these informal channels. Frequent seminars, meetings, and conferences provide opportunities for bureaucrats, politicians, and sector representatives to network and hear presentations and debates on the latest policy issues. Ministers get a chance to make important policy announcements to a mostly friendly audience and private-sector representatives have an opportunity to put forward their concerns and recommendations to politicians and government officials. New Zealand has developed only a few exclusive off-stage settings where ministers, officials, and company representatives can discuss policy issues off the record, such as the bi-monthly Petroleum Club.

The government's policy to expand extractive industry has required an accompanying expansion of the state bureaucracy, and organized surveillance of public discourse around oil and gas development. Initially the New

Zealand public service lacked enough officials with sufficient knowledge and experience of the oil and gas industry. This has given the petroleum industry opportunities to influence government policy directly by helping draft policies and regulations and indirectly through revolving door arrangements.

STATE MANOEUVERS TO EXPEDITE OIL AND GAS DEVELOPMENT AND PROMOTE THE INDUSTRY

New Zealand has a mixed history of government involvement in national economic development, from direct intervention during the Think Big era of the 1970s to more of an arm's length incentivizing and facilitation role following the radical political reforms of the 1980s. The focus of successive governments on "growing the economy," with varying policy initiatives to ameliorate or justify the cumulative effects of development on the natural environment and Kiwis' enjoyment of it, has fueled an emergent national culture of popular environmentalism, pride in the country's clean-green image, and willingness to take action to counteract perceived threats to that environment. Prominent examples have been the Save Manapouri campaign, protests over amendments to section 4 of the Crown Minerals Act, anti-oil industry protests, and climate change marches. This turbulent political economic history dotted with populist environmental "tipping points" helps explain at least in part why the National government under Prime Minister John Key has eschewed a heavy-handed interventionist approach to expanding the petroleum industry in order to avoid criticisms of "growth at all costs." The primary manoeuvers under the government's light-handed approach have been legislative reform, "factual" information dissemination, official propaganda and ministerial cheerleading (e.g., the benefits of royalties and jobs), agenda-driven funding, and diminishing community power mostly through legislative change. The crux of the struggle between the state/oil industry coalition and environmental organizations and community anti-oil opponents has been the Resource Management Act (RMA), which the government and its supporters (developers, farmers, big business, and the petroleum industry) view as the principal impediment to economic growth.

In his recent volume of case studies Bebbington[4] found a convergence of government strategies regarding extractive industry development among several South American countries. The main elements shared among these strategies were legislative reforms to facilitate expansion of extractive industry, ownership, and tax changes to increase state revenues for social investment; and "political and discursive practices that suggest *an increasingly intolerant attitude to protest and debate*." The elements have a striking resonance with New Zealand's recent experience. The National government has placed a priority on legislative reform, tax incentives, and public propaganda and debate, while seeking greater collaboration and dialogue with mainstream environmentalists and marginalizing and denigrating local anti-petroleum protest groups and Māori activists.

Petroleum exploration and production companies operating in New Zealand have influenced state policies and responses to attacks on the industry in at least three ways. First, by bringing indirect pressure to bear on the government by releasing media statements, holding interviews, and making public presentations at conferences and symposia highlighting industry needs and concerns, and criticizing government laxness in addressing these. Secondly, individual corporations and trade association PEPANZ have lobbied key ministers and officials through submission of position papers, consultations and confidential discussions (e.g., the petroleum representatives' meeting with Minister of Energy and Resources Simon Bridges that led to legislation banning protests in proximity to offshore drilling operations). And thirdly, influence has been exerted through coordinated PR, promotional visits to the regions, and confidential meetings with local authorities to "manage" community participation in resource consents and protest activities.

ENVIRONMENTALIST OPPOSITION, COMMUNITY PROTEST, AND INSTITUTIONAL CHANGE

This study has argued that the development of closer collaborative relations between the New Zealand state and the petroleum industry, and the strategies each has adopted to promote and defend the industry have resulted in a significant shift in long-standing government/community sector relations. The move toward closer state/Big Oil collaboration in turn has been influenced by at least four important developments: (1) globalization, deregulation, and opening the country's natural resources to multinational corporate expropriation; (2) a high government debt level and the need to boost export earnings; (3) growing public concern over National government efforts to expedite onshore and deep-sea oil and gas expansion while prevaricating on climate change; and (4) increased lobbying, counter-propaganda, and protest action by established environmental organizations, anti-fossil fuel groups, and Māori activists as well as spontaneous local protests by concerned citizens' groups.

The international literature on the contentious relationship between extractive industry and local communities and indigenous peoples suggests that corporations are less likely to contribute to authentic *sustainable development*, more able to manipulate community engagement exercises to suit their own ends, and less likely to prevent or compensate for the damage they cause in countries that have weak social institutions, poor regulation and opaque, dysfunctional democracies.[5] New Zealand enjoys relatively transparent governance, strong institutions, and, as the Minister of Energy and Natural Resources is proud of declaring, a "world class regulatory environment." Nevertheless, environmental organizations, community anti-petroleum activists, and even some municipal councils have criticized the lack of government and industry consultation, decreased control over regional development, growing environmental degradation, and inadequate sharing of revenues that could offset social and infrastructure costs. Thus there must be other processes at work that allow extractive corporations to extend their influence and affect state/community sector relations, even where strong institutions and an open democracy exist.

The answer, I suggest, can be found in the evolving relationships among key resource management and development planning institutions at national and local levels, and the factors contributing to the transformation of these relations. In terms of our initial regulation diagram (chapter 1), in New Zealand there has been (a) a strengthening of the two-way relationship between government and the oil industry, (b) a weakening of the relationship between the community sector and the extractive sector, and (c) an increased dominance (one-way) of the community and local government sector by central government motivated by concerns to remove obstructions to accelerated economic development. The pattern of these changing relations can be seen in the following diagram (figure 7.1).

As noted in chapter 1, it is not unusual for "frontier" countries experiencing a resource extraction boom or where the state is working with the extractive sector to rapidly expand mining and drilling to undergo a dramatic transformation in their regulatory institutional configuration over a short period of time. The result can be expanded influence and potential rent-seeking on the part of extractive corporations, more authoritarian behavior by the state in relation to civil society, and increasing intolerance to protest and debate. Ross found that resource rich governments tend to become more authoritarian, use low tax rates and patronage to dampen demands for greater democracy, and strengthen legislation and tighten security to repress troublesome NGOs and popular movements.[6] Even moderate regimes tend to become less consultative, either coopting or subverting civil sector NGOs, environmental groups, and community organizations.

This tendency has become apparent in New Zealand as the government has pushed to greatly expand oil and gas exploration. At a structural level, such developments are reflected in historically significant transformations that have occurred in institutional arrangements which have linked the state and

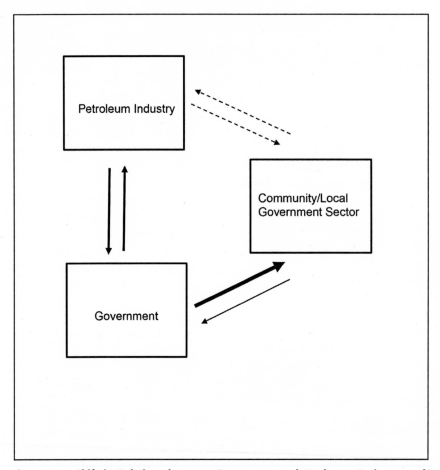

**Figure 7.1. Shift in Relations between Government and Moderate Environmental/
Recreational Groups.**

the community/local government sector for decades. As noted previously,
they have been brought on by the government's efforts to boost export-led
growth by encouraging expansion of dairying and extractive industry.

Two clusters of institutional relationships are involved, one formal and
the other informal. The *first cluster* has to do with national development,
natural resource management, regional development, and local government/
community planning. This institutional cluster includes (a) government
structures and procedures for establishing national development priorities
and developing resource management and safety policies, legislation, regulations, monitoring, and enforcement procedures; (b) environment courts and
regulatory agencies; (c) extractive industry environmental impact planning

and community engagement policies; (d) regional council resource planning and monitoring systems; (e) local resource consenting and appeal processes; and (f) local council long-term planning, and local structures and processes for planning and implementing economic development.

This cluster of institutions has been informed by principles and priorities set out in resource management, conservation, and community planning legislation, as well as various Māori development and community development policies and programs. All of these have traditionally placed a high value on *environmental preservation* and on *public participation, input, and consultation* which (it has been tacitly understood) the government of the day would take into account when formulating policy.

Over the past quarter-century, successive governments have sought to transfer more responsibility to local councils, communities, and Māori iwi for managing their own affairs. During the 1990s, National governments borrowed from Reaganomics and stressed subsidiarity: i.e., decisions should be taken at the lowest possible level closest to where they will have their greatest effect. The Labour government under Helen Clark borrowed from Tony Blair's "Middle Way," emphasizing a sustainable development approach to national and local development and undertaking initiatives to empower communities and build Māori capacity for self-determined development. As a consequence local authorities, community groups, and Māori tribal organizations came to expect they would be given greater responsibility and central government backing for local planning, development, and stewardship of natural resources.

When the National government assumed office in 2008 and introduced the Business Growth Agenda to accelerate commodity-led export growth and expand petroleum development, it initiated a campaign of reform, promotion, and defense that fundamentally altered the local planning, resource management, and environmental protection institutions and values that had evolved during the previous quarter-century. The effect was to shift the balance of power back toward central government, "in the National interest" of course. National maintained, for instance, that local government long-term planning for sustainable development, the extensive involvement of ordinary citizens, and contestable resource consent processes were impediments to, rather than essential components of, national development.

The National government's campaign of legislative reform and oil and gas expansion proceeded by challenging the validity and efficacy of some of the values and priorities that have guided the functioning of these important state/ community sector institutions. Their efforts were spurred by increasing environmentalist criticism and sporadic grassroots protests regarding their policies. By the end of 2015, the government felt it had sufficient backing from rural and business sectors to push ahead with more radical reforms. These more hardline measures resulted in further changes in institutional relations between the state and the community/local government sector. The Productivity Commission's blue-skies review, touted as being widely consultative, was predicated on getting participants to buy in to rethinking some of these same values and priorities. Except that this time the rethinking was being driven by the state rather than civil society, as was the case with the original RMA. Similar pressures were placed on existing institutional relations as a consequence of Environment Minister Nick Smith's the 2015 Resource Legislation Amendment Bill, which critics including the Parliamentary Commissioner for the Environment pointed out would further centralize decision-making and unduly increase ministerial powers to intervene in local government environmental consenting processes, over-ride regional plans, and restrict citizen participation in consent hearings and appeals. At the same time, Economic Development Minister Steven Joyce directed his ministry officials to work with local development agencies and business sectors to formulate Regional Action Plans, while excluding local councils and the public for fear they would slow down the planning process and introduce extraneous social and environmental concerns.

Meanwhile, calls for an "ecological revolution" in policy-making to place greater emphasis on valuing nature and urgently developing a transition plan to a clean energy economy for the most part fell on deaf ears. The government chose to focus on "sustainable growth" and, after carefully stage-managed consultation, would eventually get around to developing an economically cautious transition plan that suited its own agenda.

Transformations have also occurred within a *less-formal cluster* of institutional relationships including regular face-to-face meetings with ministers, invited consultations on policy issues, technical advisory groups, and multistakeholder conferences, and working groups which successive governments have used to engage and consult behind the scenes with community sector representatives, in this case more moderate environmental and recreation advocates. As Rudel, Roberts, and Carmin note (chapter 1), "States have typically created corporatist policymaking circles that include long-established, moderate environmental nongovernmental organizations and exclude disadvantaged and unorganized peoples." Under New Zealand's National government, relevant ministers have made efforts to consciously move relations with such groups from confrontation to greater collaboration, while continuing to attack and dismiss extremist environmental groups and anti-oil protesters. The Environment Minister's proposal in the Resource Legislation Amendment Bill 2015 to replace adversarial resource consent hearings with a more collaborative process involving hand-picked local stakeholder representatives is a case in point. These informal collaborative-appearing institutions are not a contradiction to government's broader efforts to dominate relations regarding regional development, resource management, and environmental protection. They are after all focused primarily on managing and mitigating the effects of accelerated extractive and agricultural development, not challenging *whether* such developments should occur in the first place. They are also useful in conveying to the general public the impression that the government is "environmentally responsible" and committed to consultation, even if it has little effect on policy particularly in regard to economic development and resource management.

The nature of changes in these institutional relations, and the values and priorities that guided them, have profound implications for the country's future development. After three terms driving the Business Growth Agenda, the government is clearly calling the shots, going through the motions of community and iwi engagement[7] but saying in effect "trust us, we know what we're doing." The broad direction of these institutional transformations has been:

- From localism back toward centralism (Oliver Hartwich in his 2013 treatise on localism observes that New Zealand's centralist government tendencies are historically rooted and extreme by international standards; but there have been occasional experiments with localism, as for instance during the last Labour government);
- From inclusive, active citizen engagement in local and regional development planning (much more than mere "consultation") to "representative," central government-driven development; and
- From "sustainable development" to "sustainable growth" (technically, from policies that support socially and environmentally optimal productive activity to economically maximizing activities that transfer externality costs to society and the environment; or what critics like Sir Geoffrey Palmer have termed "development at all costs").

All three transformations, I would argue, are direct manifestations of accelerated globalization: the growing influence of transnational corporations, pressures on states to collaborate with the expansion of international capital, and the declining ability of regions and local communities to exercise control over their own development.

The impetus for government exerting greater influence over these clusters of state/civil society institutions and closer petroleum industry collaboration has come largely (though not exclusively) from a loose coterie of "hardened" environmental organizations, anti-oil/anti-fracking groups, and Māori activists who have attacked government's contradictory energy policies, disrupted industry conferences, engaged in civil disobedience against petroleum companies, kept up a campaign of resource consent submissions, and exposed shoddy regulatory practices and environmentally damaging company operations. From the government's perspective (and that of the petroleum industry), these groups are quite literally beyond the pale of institutional participation and policy-making. They are accused of having an agenda, using experts, being well-funded, and being well-organized (not unlike fossil fuel industry associations and pro-oil lobby groups). The result has been a widening distinction in the eyes of the state and petroleum companies, between more reasonable, moderate environmental and recreational organizations they are prepared to deal with and extremist elements who refuse to engage unless the government changes its policies and the "energy" industry transitions to renewables. Interestingly, from a Regulationist perspective, New Zealand's configuration of institutional relations regarding natural resource management and national development is beginning to show some of the characteristics of the resource curse.

Senior National ministers are experienced political strategists and are no doubt conscious of the risks to the government's Business Growth Agenda from adopting a more centrally driven rather than facilitative approach, weakening long-standing institutions of community-based planning and citizen engagement, and denigrating environmental/recreational and anti-fossil fuel "extremists." So far the government has been able to rely on the acquiescence if not active support of Middle New Zealand. But several factors could significantly undermine the government's current position. These include (1) the effects of a slumping global oil market on oil and gas exploration in New Zealand, damaging the much vaunted Taranaki economy and raising questions over the veracity of the government's hype about the benefits of oil and gas development; (2) rising public concerns over climate change; and (3) the country's history of popular environmentalist "uprisings" in response to major threats to the environment, exposure of industry harm and/or environmentally compromising state policies. As one environmental activist I interviewed stated "when our political institutions fail to address these issues, direct action is often the only option we have left." The combination of these threats to the government's "sustainable growth" agenda could reach a tipping point as soon as the next election if recent polls in the United States, Australia, and New Zealand are anything to go by.[8] Even if National were elected to a fourth term, such a tipping point and the social conflict that could accompany it may still eventuate if the government were to proceed with further radical reforms of the Resource Management Act, persist in its contradictory energy policy, and continue to procrastinate in developing a comprehensive strategy for transitioning to a low-carbon future.

NOTES

1. Bebbington, ed., *Social Conflict*; Kirsch, "Afterword" in Bebbington, ed., *Social Conflict*. Kirsch, *Mining Capitalism*. Benson and Kirsch, "Capitalism and Politics."

2. Upstream E&P companies virtually never advertise in the media, and any publically identified sponsorship of sporting teams and events is low-key.

3. Benson and Kirsch, "Capitalism and the Politics of Resignation," 2010, 8–9.

4. Bebbington, ed, *Social Conflict*, 214ff.

5. E.g., Eduardo Canel, Uwafiokun Idemudia and Liisa North, "Rethinking Extractive Industry: Regulation, Dispossession and Emerging Claims," *Canadian Journal of Development Studies/Revue Canadienne d'Etudes du Developpement*, Volume 30, 1-2, 2010: 5–25. doi:10.1080/02255189.2010.9669279. Stuart Kirsch, "Sustainable Mining," *Dialectical Anthropology*, 34, 1 (2010): 87–93.

6. Michael Ross. "Does Oil Hinder Democracy?" *World Politics* 53, 3 (April 2001), 325–361. See also Jacqueline DeMeritt and Joseph Young, "A Political Economy of Human Rights: Oil, Natural Gas, and State Incentives to Repress," *Conflict Management and Peace Science* 30, 2 (April 2013): 99–120.

7. In 2011 the government signed the Kia Tutahi Relationship Accord with community sector representatives, setting out principles and expectations for how the government and community organizations should work together regarding social services and community development initiatives. The accord replicated a similar national consultation exercise by the first Clark Labour government in 2002 that resulted in a working together agreement.

8. "Climate Change a Vote-Changer at Federal Election, Says Poll," *The Guardian*, March 17, 2016; "A Tipping Point': Record Number of Americans See Global Warming as Threat," *The Guardian*, March 18 2016. A 2012 survey sponsored by the New Zealand mining and petroleum industries found that a majority of respondents did not object to developing coal and petroleum resources as long *as "the environment is protected" (79 percent) and there was "no increase in greenhouse emissions" (71 percent).* Neither proviso was going to be met, as most survey respondents would no doubt have been aware, but that didn't stop the trade associations claiming Kiwis were not opposed to mining and drilling.

Glossary of Māori Terms

Aotearoa: Maori name for New Zealand, sometime translated "land of the long white cloud"

Aotearoa/New Zealand: alternative name for New Zealand; Maori is an official language of the country

Hapū: kinship group, clan, subtribe

Hīkoi: march, hike, tramp

Hui: assembly, gathering

Iwi: extended kinship group, tribe, nation, people

Kaitiakitanga: guardianship, stewardship, trusteeship

Kanohi ki te kanohi: face to face, in person

Kia Tūtahi: standing together

Kuia: elderly woman, grandmother, female elder

Mana: prestige, authority, control, power

Māori: indigenous inhabitant of Aotearoa/New Zealand

Marae: courtyard in front of the *wharenui* where formal greetings and discussions are held; the complex of buildings around the marae

Pākehā: New Zealander of European descent

Papatūānuku: Mother Earth, the Land

Pōwhiri: welcome ceremony on a marae, ritual of encounter

Rohe: territory, region, area of land

Rūnanga: council, tribal council, board (as in business)

Tairāwhiti: customary Māori name for the Gisborne region

Tangata whenua: local people, Maori, hosts, indigenous inhabitants of a place

Tapu: sacred, holy, prohibited

Tūrangawaewae: place to stand, location to which one is connected and feels empowered

Wānanga: seminar, conference, forum, center of learning
Whānau: family group, extended family
Wharenui: meeting house, large house

Bibliography

Adams, Amy. "Improving New Zealand's Resource Management System." Media release, New Zealand Government, Wellington, March 1, 2013. https://www.bee hive.govt.nz/minister/amy-adams.

Adams, Amy. "New Regulations for Exploratory Drilling." Media release, New Zealand Government, Wellington, February 27, 2014. https://www.beehive.govt.nz/release/new-regulations-exploratory-drilling.

Adams, Michael. "East Coast North Island – Oil Resource Play – Development Scenario Models." Report to the Ministry of Business, Innovation and Employment, New Zealand Government, Wellington, October 2012. http://www.mbie.govt.nz/info-services/sectors-industries/natural-resources/oil-and-gas/petroleum-expert-reports/east-coast-oil-and-gas-development-study/MARE-EC-Resource-Plays-scenarios-East-Coast.pdf.

Alinsky, Saul. *Rules for Radicals.* London: Vintage Books, 1971/1989.

Altman, Jon, and David Martin eds. *Power, Culture, Economy: Indigenous Australians and Mining.* Research Monograph 30, Centre for Aboriginal Economic Research, Canberra, Australian National University, 2009.

Arellano-Yanguas, Javier. "Mining and Conflict in Peru: Sowing the Minerals, Reaping a Hail of Stones." In *Social Conflict, Economic Development and Extractive Industry: Evidence from Latin America,* edited by Anthony Bebbington, 89–111. London: Routledge, 2012.

Argent, Nathan, and Simon Boxer. *The Future is Here: New Jobs, New Prosperity and a New Clean Economy.* Greenpeace New Zealand, Auckland, 2013. http://www.greenpeace.org/new-zealand/Global/new-zealand/P3/publications/climate/2013/TheFutureisHereGREENPEACEreport.pdf.

Argent, Nathan. "Our Right to Protest Is Vital." Greenpeace New Zealand, Annual Impact Report, Auckland, 2013. http://www.greenpeace.org/new-zealand/Global/new-zealand/P3/publications/other/Greenpeace%20NZ%20Annual%20Impact%20Report%202013.pdf.

Baker, C. D., and J. B. Napp. "Minerals Briefing Paper 2014: Policies for Increasing New Zealand's Attractiveness for Investment in Responsible Minerals Exploration and Mining." Straterra Inc, Wellington, April 2014. www.straterra.co.nz.

Baker Hughes Corporation. "Hydraulic Fracturing: An Environmentally Responsible Technology for Ensuring Our Energy Future." White paper, 2012. Accessed March 3, 2015. http://www.bakerhughes.com/news-and-media/resources/white-papers/hydraulic-fracturing-white-paper.

Banks, Glenn. "Activities of TNCs in Extractive Industries in Asia and the Pacific: Implications for Development." In *Transnational Corporations* 18, 1 (April 2009), United Nations, Geneva: 51–68.

Barth, Janette. "Critique of PPI Study on Shale Gas Job Creation." January 2, 2012. Accessed February 28, 2013. Catskill Mountainkeeper, http://www.catskillcitizens.org/barth/JMB_Critique_of_PPI_Jan_2_2012.pdf.

Bateman, Christopher. "A Colossal Fracking Mess." *Vanity Fair*, June 2010. http://www.vanityfair.com/news/2010/06/fracking-in-pennsylvania-201006.

Beatson, Andrew. "Summary of Consultation between TAG Oil and Iwi in Relation to Proposed Exploration Well Known as Punawai-1," Gisborne District Council, letter to Hans van Kregten, Environment and Policy Group Manager on July 8, 2013. file:///C:/Users/Terrence/Downloads/DOCSn311850v1BellGullyIwiConsultationletter.pdf.

Bebbington, Anthony, ed. *Social Conflict, Economic Development and Extractive Industry: Evidence from Latin America.* London: Routledge ISS Studies in Rural Livelihoods, 2012.

Bebbington, Anthony. "Extractive Industries, Socio-environmental Conflicts and Political Economic Transformations in Andean America." In *Social Conflict, Economic Development and Extractive Industry: Evidence from Latin America*, edited by Anthony Bebbington, 1–26. London: Routledge, 2012.

Bell Gully. "Draft Regulations for Exploratory Oil and Gas Drilling under the EEZ Act." *Lexology*. December 19, 2013.

Benson, Peter, and Stuart Kirsch. "Capitalism and the Politics of Resignation." *Current Anthropology* 51, 4 (2010): 459–486.

Benson, Peter, and Stuart Kirsch. "Corporate Oxymorons." *Dialectical Anthropology* 34 (2010): 45–48. doi 10.1007/s10624-009-9112-y.

Berman, Arthur. "U.S. Shale Gas: Magical Thinking and the Denial of Uncertainty." Presentation at a workshop on Environmental and Social Implications of Hydraulic Fracturing and Gas Drilling in the United States: An Integrative Workshop for the Evaluation of the State of Science and Policy, Nicholas School of the Environment, Duke University, Durham, NC, January 9, 2012. Similar presentation at the James A Baker III Institute for Public Policy, Rice University, Houston, TX, January 18, 2012. https://www.google.co.nz/url?sa=t&rct=j&q=&esrc=s&source=web&cd=1&cad=rja&uact=8&ved=0ahUKEwjH6I-x_-nNAhWFmpQKHRudANkQFggaMAA&url=https%3A%2F%2Fbakerinstitute.org%2Ffiles%2F2381%2F&usg=AFQjCNGYjQrEXSVNYWIqYu_QrCrRNcvbkQ.

Berman, Arthur. "IEA Offers No Hope for an Oil-Price Recovery." *Petroleum Truth Report*, November 13, 2015. Accessed November 26, 2015.

Berman, Richard. "Big Green Radicals: Exposing Environmental Groups." Presentation to the Western Energy Alliance Annual Meeting, Colorado Springs, June 20, 2014.

Bertram, Geoff. "Mining in the New Zealand Economy." *Policy Quarterly* 7, 1 (2011): 13–19.

Bertram, Geoff. "Lessons from Think Big." Presentation to the New Zealand Association of Impact Assessment conference on Assessing the Impacts of Petroleum and Mineral Extraction in New Zealand, Wellington, December 10–11, 2012.

Bourdieu, Pierre. *Outline of a Theory of Practice*, Translated by Richard Nice. Cambridge: Cambridge University Press, 1977.

Brasier, Kathryn, Matthew Filteau, Diane McLaughlin, Jeffrey Jacquet, Richard Stedman, Timothy Kelsey, and Stephan Goetz. "Residents' Perceptions of Community and Environmental Impacts from Development of Natural Gas in the Marcellus Shale: A Comparison of Pennsylvania and New York Cases." *Journal of Rural Social Sciences* 26, 1 (2011): 32–61.

Bridges, Simon. "Speech to Advantage NZ: 2013 Petroleum Conference." Media release: New Zealand Government, Wellington, April 29, 2013. https://www.bee hive.govt.nz/speech/speech-advantage-nz-2013-petroleum-conference.

Bridges, Simon. "Block Offer 2014: Address to the Advantage NZ 2014 Geotechnical Petroleum Forum." New Zealand Government, Wellington, April 2, 2014. https://www.beehive.govt.nz/speech/block-offer-2014-address-advantage-nz-2014 -geotechnical-petroleum-forum.

Bridges, Simon. "Speech to 2014 New Zealand Petroleum Summit." Media release: New Zealand Government, Wellington, October 1, 2014. https://www.beehive .govt.nz/speech/speech-2014-new-zealand-petroleum-summit.

Broomfield, Mark. *Support to the Identification of Potential Risks for the Environment and Human Health Arising from Hydrocarbons Operations Involving Hydraulic Fracturing in Europe.* Report to the European Commission Directorate-General Environment, AEG Plc, ED57281 Issue Number 17c. Didcot, UK. 10 August 2012. Accessed February 12, 2015, http://ec.europa.eu/environment/integration/energy/pdf/fracking%20study.pdf.

Brownlee, Gerry. "Unlocking New Zealand's Petroleum Potential." Media release, New Zealand Government, Wellington, November 18, 2009. https://www.beehive .govt.nz/release/unlocking-new-zealand039s-petroleum-potential.

Brownlee, Gerry, and Kate Wilkinson. "Time to Discuss Maximising our Mineral Potential." Media release, New Zealand Government, Wellington, March 22, 2010. https://www.beehive.govt.nz/release/time-discuss-maximising-our-mineral -potential.

Bush, Chris. "Chairman's Address: Our Petroleum Future." Address to the New Zealand Petroleum Summit, Wellington, September 19, 2012.

Campbell, Karen, and Matt Horne. "Shale Gas in British Columbia: Risks to BC's Water Resources." The Pembina Institute, Calgary, 2011. Accessed May 29, 2013. https://www.pembina.org/reports/shale-and-water.pdf.

Canel, Eduardo, Uwafiokun Idemudia, and Liisa North. "Rethinking Extractive Industry: Regulation, Dispossession and Emerging Claims." *Canadian Journal of Development Studies/Revue Canadienne d'Etudes du Developpement* 30, 1–2 (2010): 5–25. doi:10.1080/02255189.2010.9669279.

Chivers, Danny. "Fracking: the Gathering Storm." *New Internationalist*. December 3, 2013. Accessed December 14, 2013.

Christopherson, Susan. "Marcellus Hydro-Fracturing: What Does it Mean for Economic Development?" Slide presentation, Department of City and Regional Planning, Cornell University, April 7, 2011. Accessed September 2, 2013, from Catskill Mountainkeeper, www.catskillmountainkeeper.org/.

Cleary, Paul. *Mine Field: The Dark Side of Australia's Resources Rush.* Collingwood, Victoria: Black Inc., 2012.

Cobb, Kurt. "Orwellian Newspeak and the Oil Industry's Fake Abundance Story." *Resilience*, July 11, 2014. Accessed October 21, 2014. www.resilience.org.

Colborn, Theo, Carol Kwiatkowski, Kim Schultz, and Mary Bachran. "Natural Gas Operations from a Public Health Perspective." *Human and Ecological Risk Assessment: An International Journal* 17, 5 (September 2011): 1039–1056.

Cooley, Heather, and Kristina Donnelly. *Hydraulic Fracturing and Water Resources: Separating the Frack from the Fiction,* Pacific Institute, Oakland, CA, 2012. Accessed February 12, 2014. http://pacinst.org/publication/hydraulic-fracturing-and-water-resources-separating-the-frack-from-the-fiction/.

Coull, David. "Petroleum Business in New Zealand: the Continuing Challenge." Presentation to the Advantage New Zealand Petroleum Conference, Auckland, April 30, 2013.

Coull, David. "Significant Period of Change and Regulatory Reform in the New Zealand E&P Sector." *Who's Who Legal*, June 2014. Accessed June 19, 2014. www.whoswholegal.com.

Darrell, Bruce. "Why Confusion Exists over When Peak Oil Will Occur." Foundation for the Economics of Sustainability (FEASTA), January 12, 2007. Accessed on June 7, 2014. http://www.feasta.org/2007/01/12/why-confusion-exists-over-when-the-oil-peak-will-occur/.

Davey, Brian. "Peak Oil Revisited . . ." Foundation for the Economics of Sustainability, June 3, 2014. Accessed June 7, 2014. http://www.feasta.org/2014/06/03/peak-oil-revisited/.

Davies, Richard J., Sam Almond, Robert S. Ward, Robert B. Jackson, Charlotte Adams, Fred Worrall, Liam G. Herringshaw, Jon G. Gluyas, and Mark A. Whitehead. "Oil and Gas Wells and Their Integrity: Implications for Shale and Unconventional Resource Exploitation." *Marine and Petroleum Geology* 56 (September 2014): 239–254.

DeMeritt, Jacqueline, and Joseph Young. "A Political Economy of Human Rights: Oil, Natural Gas, and State Incentives to Repress." *Conflict Management and Peace Science* 30, 2 (April 2013): 99–120.

Department of Internal Affairs. *Smarter Government, Stronger Communities: Towards Better Local Governance and Public Services.* New Zealand Government, Wellington, February 2, 2011. https://www.dia.govt.nz/Resource-material-Our-Policy-Advice-Areas-Smarter-Government-Stronger-Communities.

Department of Internal Affairs. *Report of the Local Government Efficiency Taskforce.* New Zealand Government, Wellington, December 2012. https://www.dia.govt .nz/pubforms.nsf/URL/Local-Government-Efficiency-Taskforce-Final-Report -11-December-2012.pdf.

Dunlop, R. E., and A. M. McCright. "Organized Climate Change Denial." In *The Oxford Handbook of Climate Change and Society*, edited by J. S. Drysek, R. B. Norgaard and D Schlosberg, 144–160. Oxford: Oxford University Press, 2011.

Diprose, G., A. C. Thomas, and S. Bond. 'It's Who We Are': Eco-nationalism and Place in Contesting Deep-sea Oil in Aotearoa New Zealand." *Kōtuitui: New Zealand Journal of Social Sciences Online* 11, 2 (2016): 159–173. doi: 10.1080/1177083X.2015.1134594

Dutzik, Tony, Elizabeth Ridlington and John Rumpler. "The Costs of Fracking: The Price Tag of Dirty Drilling's Environmental Damage." Environment New York Research & Policy Center (Fall 2012). Accessed February 1, 2013. www .environmentnewyorkcenter.org/.

Ellsworth, William. "Injection-induced Earthquakes." *Science* 341, 6142 (July 12, 2013). doi: 10.1126/science.1225942.

Engdahl, William F. "America: The New Saudi Arabia?" *Voltaire Network.* March 15, 2013. Accessed June 3, 2013.

Environmental Defence Society. "EDS Critical of Proposed RMA Reforms." Newsletter, August 10, 2013. Accessed September 13, 2013. www.eds.org.nz.

Environmental Defence Society, Report by DLA Phillips Fox, "Implications of Proposed Changes to Part 2 of the RMA." September 2, 2013, para 9, www.eds .org.nz.

Environmental Defence Society. "EDS Position Paper on RMA Reforms." Newsletter, September 3, 2013. Accessed September 13, 2013. www.eds.org.nz.

Environmental Protection Agency. "Hydraulic Fracturing for Oil and Gas and Its Potential Impact on Drinking Water Resources." Draft report, EPA, Washington, DC, June 2015. http://www.epa.gov/hfstudy/study-potential-impacts-hydraulic -fracturing-drinking-water-resources-progress-report-0.

Epstein, Anne C. "Health Risks of Living near Natural Gas Development." Slide presentation to the 100% Renewable Denton Town Meeting, University of North Texas, Denton, Texas. Accessed April 4, 2016. http://www.scribd.com/ doc/306021913/Health-Risks-of-Living-Near-Natural-Gas-Development.

Finer, M., C. Jenkins, S. Pimm, B. Keane, and C. Ross, 2008. "Oil and Gas Projects in the Western Amazon: Threats to Wilderness, Biodiversity and Indigenous Peoples." *Plos one* 3, 8 (2008): 1–9.

Fish and Game New Zealand, "Fish and Game Opposes 'Attack' on RMA." Media release, Wellington, September 11, 2012. http://www.fishandgame.org.nz/newsitem/ fish-game-opposes-attack-rma.

Food and Water Watch. "Exposing the Oil and Gas Industry's False Jobs Promise for Shale Gas." Washington, DC, 2011. Accessed July 21, 2014. http://www.foodand waterwatch.org/insight/exposing-oil-and-gas-industrys-false-jobs-promise-shale -gas-development.

Food and Water Watch. "Fracking Colorado: Illusory Benefits, Hidden Cost." Issue
    Brief, Denver, August 2013. Accessed July 21, 2014. http://www.foodandwater
    watch.org/insight/fracking-colorado.
Franks, Daniel, David Brereton, and Chris Moran. "Managing the Cumulative Im-
    pacts of Coal Mining on Regional Communities and Environments in Australia."
    Impact Assessment and Project Appraisal 28, 4 (December 2010): 299–312. doi:
    10.3152/146155110X12838715793129.
Franks, Daniel, and Rachel Davis. "The Costs of Conflicts with Local Communities
    in the Extractive Industry." Paper presented to the First International Seminar on
    Social Responsibility in Mining, Santiago, Chile, October 19–21, 2011.
Franks, Daniel, Jo-Anne Everingham, and David Brereton. "Governance Strategies
    to Manage and Monitor Cumulative Impacts at the Local and Regional Level."
    ACARP Project C19025. Centre for Social Responsibility in Mining, University of
    Queensland, Brisbane, 2012.
Freudenburg, William. "Women and Men in an Energy Boomtown: Adjustment,
    Alienation, Adaptation." *Rural Sociology* 46:2 (1981): 220–244.
Freudenberg, William, and Robert Wilkinson eds. *Equity and the Environment*. Re-
    search in Social Problems and Public Policy 15. Amsterdam: Elsevier 2008.
Freyman, Monika. "Hydraulic Fracturing and Water Stress: Water Demand by the
    Numbers." Ceres, 2014. Accessed June 1, 2015. http://www.ceres.org/resources/
    reports/hydraulic-fracturing-water-stress-water-demand-by-the-numbers.
Frynas, Jedrzej George. "The False Developmental Promise of CSR: Evidence from
    Multinational Oil Companies." *International Affairs* 81, 3 (May 2005): 581–598.
Gedricks, Al. *Resource Rebels: Native Challenges to Mining and Oil Companies*.
    Cambridge, MA: South End Press, 2001.
Giddens, Anthony. *Runaway World: How Globalization is Reshaping Our Lives*.
    London: Profile Books, 2002.
Gilberthorpe, Emma, and Glenn Banks. "Development on Whose Terms? CSR Dis-
    course and Social Realities in Papua New Guinea's Extractive Industries Sector."
    Resources Policy 37, 2 (June 2012): 185–193.
Gilmore, J. S. "Boom Towns May Hinder Energy Resource Development: Isolated
    Rural Communities Cannot Handle Sudden Industrialization and Growth without
    Help." *Science* 191 (1976): 535–540.
Godelier, Maurice. *Rationality and Irrationality in Economics*. New York: Monthly
    Review Press, 1972.
Goldman, Gretchen, 2013. "Colorado Towns Pass Fracking Moratoria, Bans De-
    spite Big Spending by the Oil and Gas Industry." *The Equation* blog. Union
    of Concerned Scientists USA. Accessed November 9, 2013. http://blog.ucsusa
    .org/gretchen-goldman/colorado-towns-pass-fracking-moratoria-bans-despite-big
    -spending-by-the-oil-and-gas-industry-293.
Goodell, Jeffrey. "The Big Fracking Bubble: The Scam behind Aubrey McClendon's
    Gas Boom." *Rolling Stone* magazine, March 1, 2012.
Greenpeace New Zealand. *The Future is Here: New Jobs, New Prosperity and a
    New Clean Economy*. Auckland, February 2013. http://www.greenpeace.org/new
    -zealand/en/campaigns/climate-change/The-Future-is-Here/.

Greenpeace New Zealand. "Energy Minister Challenged to Release Full Meeting Details over Misleading Parliament Accusation." Media release, June 4, 2013. http://www.greenpeace.org/new-zealand/en/press/Energy-Minister-Challenged-to -Release-Full-Meeting-Details-Over-Misleading-Parliament-Accusation/

Greer, Glen, Michelle Marque, and Caroline Saunders. "Petroleum Exploration and Extraction Study." Report to the Gisborne District Council. Agribusiness and Economics Research Unit, Lincoln University, 2013. Accessed May 9, 2013. www.gdc.govt.nz/assets/Files/.../Envirolinkpetroleumimpactsstudy FINAL24Jan13.pdf

Grey, Sandra, and Charles Sedgwick. "The Contract State and Constrained Democracy: the Community and Voluntary Sector under Threat." *Policy Quarterly* 9, 3 (August 2013): 3–10.

Haalboom, Bethany. "Confronting Risk: A Case Study of Aboriginal Peoples' Participation in Environmental Governance of Uranium Mining, Saskatchewan." *The Canadian Geographer / Le Géographe Canadien* 4, 58 (3) (2014): 276–290. doi: 10.1111/cag.12086.

Hager, Nicky. *Dirty Politics: How Attack Politics is Poisoning New Zealand's Political Environment.* Nelson: Craig Potton Publishing, 2014.

Hager, Nicky, and Bob Burton. "Secrets and Lies: How Shandwick PR Tried to Destroy the Rainforests of New Zealand." *PR Watch* 7, 1 (2000): 1–5.

Harmon, Richard. "The Generation X Minister Driving NZ (and His Own Career) Along the Technology Curve." *Politik*, March 3, 2016. Accessed March 5, 2016. http://politik.co.nz/en/content/politics/565/.

Hartwich, Oliver. "A Global Perspective on Localism." Published jointly by Local Government New Zealand and The New Zealand Initiative, Wellington, 2013. http://www.lgnz.co.nz/assets/Publications/A-global-perspective-on-localism.pdf.

Heinberg, Richard. *The End of Growth: Adapting to Our New Economic Reality.* Gabriola Island, BC: New Society Publishers, 2011.

Heinberg, Richard. *Snake Oil: How Fracking's False Promise of Plenty Imperils Our Future.* Santa Rosa: Post-Carbon Institute, 2013.

Henderson, Dean. "Energy Vampires." *Left Hook* (blog), January 9, 2014. Accessed March 24, 2014. www.deanhenderson.wordpress.com.

Herman, Edward, and Noam Chomsky. *Manufacturing Consent: the Political Economy of the Mass Media.* London: The Bodley Head, 2008.

Hertsgaard, Mark. "Scientists Warn: The Paris Climate Agreement Needs Massive Improvement." *The Nation*, December 11, 2015.

Hide, Rodney. "Local Government Act 2002 Amendment Act: Decisions for Better Transparency, Accountability and Financial Management of Local Government." Department of Internal Affairs, Wellington, November 26, 2010. http://www.bay buzz.co.nz/wp-content/uploads/2010/08/HideLGA.pdf.

Hilson, Gavin. "Corporate Social Responsibility in the Extractive Industries: Experiences from Developing Countries." *Resources Policy,* Special Issue 37, 2 (2012): 131–137. http://dx.doi.org/10.1016/j.resourpol.2012.01.002.

Hollingsworth, J. Rogers, and Robert Boyer, eds. *Capitalism: The Embeddedness of Institutions.* Cambridge: Cambridge University Press, 1997.

Howarth, Robert. "Methane and the Greenhouse Gas Footprint of Shale Gas." Presentation to the 100% Renewable Denton Town Hall Meeting, University of North Texas, Denton, Texas, March 25, 2016. Accessed 4 April, 2016. http://www.scribd.com/doc/306019583/Methane-and-the-Greenhouse-Gas-Footprint-of-Shale-Gas.

Howarth, Robert, Renee Santoro, and Anthony Ingraffea. "Methane and the Green-house-Gas Footprint of Natural Gas from Shale Formations." *Climatic Change* 106 (2011): 679–690. Accessed September 7, 2014. doi 10.1007/s10584-011-0061-5.

Hughes, David. *Drill Baby Drill: Can Unconventional Fuels Usher in a New Era of Energy Abundance?* Santa Rosa: Post Carbon Institute, 2013.

Ingraffea, Anthony, Martin Wells, Renee Santoro, and Seth Shonkoff. "Assessment and Risk Analysis of Casing and Cement Impairment in Oil and Gas Wells in Pennsylvania, 2000–2012." *Proceedings of the Academy of Sciences of the United States of America* 111, 30 (July 29, 2014): 10955–10960.

International Energy Agency. "Unconventional Oil Revolution to Spread By 2019." IEA, Paris: June 17, 2014.

"International Energy Statistics." Energy Information Administration, Washington, DC, 2015. http://www.eia.gov/cfapps/ipdbproject/.

IPCC. "Climate Change 2014: Synthesis Report." Contribution of Working Groups I, II and III to the Fifth Assessment Report of the Intergovernmental Panel on Climate Change [Core Writing Team, R. K. Pachauri and L. A. Meyer (eds.)]. IPCC, Geneva, Switzerland (2014), 151 pp. Accessed July 16, 2015. http://ar5-syr.ipcc.ch/.

Jackson, Robert. "Fracking: What We Know and Don't Know about Its Impacts on Water." Presentation to the Moos Family Speaker Series on Water Resources, College of Biological Sciences, University of Minnesota, January 30, 2014. Accessed April 4, 2014. http://mediasite.uvs.umn.edu/Mediasite/Viewer/?peid=e65ac4a5549e43deb76e1e1ef27a2a2a.

Jacquet, Jeffrey. "Energy Boomtowns & Natural Gas: Implications for Marcellus Shale Local Governments and Rural Communities." Rural Development Paper 43. The Northeast Regional Center for Rural Development, Pennsylvania State University, 2009. Accessed May 23, 2013. www.nercrd.psu.edu.

Jenkins, Heledd. "Corporate Social Responsibility and the Mining Industry: Conflicts and Constructs." *Corporate Social Responsibility and Environmental Management* 2, 1 (2004): 23–34.

Jenkins, Heledd, and Natalia Yakovleva. "Corporate Social Responsibility in the Mining Industry: Exploring Trends in Social and Environmental Disclosure." *Journal of Cleaner Production* 14, 3–4 (2006): 271–284.

Joyce, Steven, and Simon Bridges. "Petroleum and Minerals Sector Boosts NZ Economy." Media release, New Zealand Government, Wellington. September 2, 2013. https://www.beehive.govt.nz/release/petroleum-amp-minerals-sector-boosts-nz-economy.

Karliner, Joshua. *The Corporate Planet: Ecology and Politics in the Age of Globalization.* San Francisco: Sierra Club Books, 1997.

Kelly, Sharon. "Fracking in Public Forests Leaves Long Trail of Damage, Struggling State Regulators (blog)," March 11, 2014. Accessed February 22, 2015. www .desmogblog.com.

Kelsey, Jane. *Reclaiming the Future: New Zealand and the Global Economy.* Wellington: Bridget Williams e-Books, 2015.

Keucker, G. "Fighting for the Forests: Grassroots Resistance to Mining in Northern Ecuador." *Latin American Perspectives* 34, 2 (2007): 94–107.

Kidner, Keely. "Beyond Greenwash: Environmental Discourses of Appropriation and Resistance." PhD diss., School of Linguistics and Applied Languages, Victoria University of Wellington, Wellington, 2015.

King, Michael. *The Penguin History of New Zealand.* Auckland: Penguin, 2003.

Kinnaman, Thomas. "The Economic Impact of Shale Gas Extraction: A Review of Existing Studies." *Ecological Economics* 70 (2011): 1243–1249. Accessed September 4, 2013 from Bucknell University Digital Commons website, http://digital commons.bucknell.edu/fac_pubs/5.

Kirsch, Stuart. "Sustainable Mining." *Dialectical Anthropology* 34, 1 (2010): 87–93.

Kirsch, Stuart. "Afterword: Extractive Conflicts Compared." In *Social Conflict, Economic Development and Extractive Industry: Evidence from Latin America,* edited by Anthony Bebbington, 201–214. London: Routledge ISS Studies in Rural Livelihoods, 2012.

Kirsch, Stuart. *Mining Capitalism: The Relationship between Corporations and Their Critics.* Oakland: University of California Press, 2014.

Kitcher, Philip. "The Climate Change Debates." *Science* 328, 4 (2010): 1230–1234. doi: 10.1126/science.1189312.

Knight, David. "Climate Change and Peak Oil: Two Sides of the Same Coin?" Paper presented to a public symposium organised by Winchester Action on Climate Change, Winchester, England, July 4, 2012, published by the Foundation for the Economics of Sustainability (FEASTA). Accessed June 7, 2014. http://www .feasta.org/2012/07/13/climate-change-and-peak-oil-two-sides-of-the-same-coin/.

Korowicz, David, 2010. *Tipping Point: Near-Term Systemic Implications of a Peak in Global Oil Production.* Paper published by the Foundation for the Economics of Sustainability (FEASTA), March 15, 2010. Accessed June 7, 2014. http://www .feasta.org/documents/risk_resilience/Tipping_Point.pdf.

Korten, David. *When Corporations Rule the World.* Bloomfield, Connecticut: Kumarian Press, 2001.

Krannich, Richard. "Social Change in Natural Resource-Based Rural Communities: The Evolution of Sociological Research and Knowledge as Influenced by William R. Freudenberg." *Journal of Environmental Studies and Sciences* 2, 1 (2012): 18–27. doi 10.1007/s13412-011-0051-y.

Krupp, Jason. *Poverty of Wealth: Why Minerals Need to be Part of the Rural Economy.* New Zealand Initiative, Wellington. December 2014. https://secure.zeald .com/site/nzinitiative/files/Poverty%20of%20Wealth%20Final%20Web.pdf.

Krupp, Jason. *From Red Tape to Green Gold.* New Zealand Initiative, Wellington. March 2015. https://secure.zeald.com/site/nzinitiative/files/Red%20Tape%20final .pdf.

Li, Fabiana. *Unearthing Conflict: Corporate Mining, Activism, and Expertise in Peru.* Durham: Duke University Press, 2015.

Live Magazine. "Is Taranaki the New Texas?" Autumn 2012: 8–14.

Loomis, Terrence. "Indigenous Populations and Sustainable Development: Building on Indigenous Approaches to Holistic, Self-determined Development." *World Development* 28, 5 (2000): 893–910.

Loomis, Terrence. "Māori Initiatives in Sustainable Development." *The Journal of Aboriginal Economic Development* 3, 2 (2003): 81–88.

Loomis, Terrence, with John Mahima. "Māori Community-based Sustainable Development: A Research Progress Report." *Journal of Development Practice* 13, 4 (2003): 399–409.

Luciani, Giancomo. "The Promise and Perils of Fracking." *International Relations and Security Network*, January 9, 2014. Accessed November 9, 2014. http://www .isn.ethz.ch/Digital-Library/Articles/Detail/?coguid=9c879a60-8a40-14e8-76c3 -2c016ae9096c&lng=en&id=175268.

McAdam, Doug and Hilary Boudet. *Putting Social Movements in their Place: Explaining Opposition to Energy Projects in the United States, 2000–2005.* Cambridge: Cambridge University Press, 2012.

McDouall, Stuart. 2010. "Schedule 4 of the Crown Minerals Act is the Perfect Time for a Rational Debate in This Country." *Q&M Magazine* 7, 3 (2010). http://contract ormag.co.nz/contractor/time-for-a-rational-debate/.

McDouall, Stuart. "Stepping Up: Options for Developing the Potential of New Zealand's Oil, Gas and Minerals Sector." Report to the Ministry for Economic Development, New Zealand Government, Wellington, June 2009. http://www.mbie.gov t.nz/info-services/sectors-industries/natural-resources/oil-and-gas/petroleum -expert-reports/McDouall-Stuart.pdf.

McElroy, Michael, and Xiu Lu. "Fracking's Future: Natural Gas, the Economy, and America's Energy Prospects." *Harvard Magazine*, January–February 2013.

McGraw, Seamus. *The End of Country: Dispatches from the Frack Zone.* New York: Random House, 2011.

McNab, K., J. Keenan, D. Brereton, J. Kim, R. Kunanayagam, and T. Blathwayt. *Beyond Voluntarism: The Changing Role of Corporate Social Investment in the Extractives Sector.* Centre for Social Responsibility in Mining, The University of Queensland, Brisbane, 2012. https://www.csrm.uq.edu.au/.

Maru, Mahanga. "NZPaM's Role and Engagement with Iwi." Presentation to the Symposium on Māori Engagement with Extractive Industry, Faculty of Law, Auckland University, June 12, 2015. http://www.law.auckland.ac.nz/en/about/events-1/ events/events-2015/01/06/symposium-on-maori-engagement-with-extractive -industry—innovati.html.

Mauro, Frank, Michael Wood, Michele Mattingly, Mark Price, Stephen Herzenberg, and Sharon Ward. "Exaggerating the Employment Impacts of Shale Drilling: How and Why." The Multi-State Research Collaborative, Penn State University, 2013. https://pennbpc.org/sites/pennbpc.org/files/MSSRC-Employment-Im-pact-11-21-2013.pdf.

Mayer, Jane. *Dark Money: The Hidden History of the Billionaires behind the Rise of the Radical Right.* New York: Doubleday, 2016.

Ministry of Business, Innovation and Employment (MBIE). "Developing Our Energy Potential: The New Zealand Energy Strategy 2011-2021, and the New Zealand Energy Efficiency and Conservation Strategy." New Zealand Government, Wellington, August 2011. http://www.mbie.govt.nz/info-services/sectors-industries/energy/energy-strategies/documents-image-library/nz-energy-strategy-lr.pdf.

MBIE. "Building Natural Resources." New Zealand Government, Wellington. December 2012. http://www.mbie.govt.nz/info-services/business/business-growth-agenda/.

MBIE. "Economic Contribution and Potential of New Zealand's Oil and Gas Industry." Occasional paper 12/07. New Zealand Government, Wellington. August 2012. http://www.mbie.govt.nz/info-services/sectors-industries/natural-resources/oil-and-gas/petroleum-expert-reports/economic-contribution-and-potential-of-new-zealand2019s-oil-and-gas-industry/.

MBIE. "East Coast Oil and Gas Development Study." New Zealand Government, Wellington, March 2013. http://www.mbie.govt.nz/info-services/sectors-industries/natural-resources/oil-and-gas/petroleum-expert-reports/east-coast-oil-and-gas-development-study/East-Coast-oil-gas-development-study-report.pdf.

MBIE. "East Coast Regional Economic Potential Stage 1: Research Review." "East Coast Regional Economic Potential Stage 2: Economic Forecasting and Transport and Skills Implications." New Zealand Government, Wellington, April 2014. http://www.mbie.govt.nz/info-services/sectors-industries/regions-cities/regional-growth-programme/east-coast.

MBIE. "Energy and Minerals Research Fund—2015 Science Investment Round Successful Proposals." Media release, September 3, 2015. https://www.beehive.govt.nz/portfolio/science-and-innovation?page=1.

Ministry for Economic Development. "New Zealand Petroleum Action Plan." New Zealand Government, Wellington, November 2009. https://www.beehive.govt.nz/release/unlocking-new-zealand039s-petroleum-potential.

Ministry for the Environment. "Technical Advisory Group 2 Report." New Zealand Government, Wellington, February 2012. https://www.mfe.govt.nz/sites/default/files/tag-rma-section6-7.pdf.

Mitchell, John. "A New Political Economy of Oil." ISSN 1062-9769, *Quarterly Review of Economics and Finance* 42, 2 (2002): 251–272.

Murphy, Tom. "Global Shale Energy: Emerging Technology, Evolving Regulations and Increasing Supply." Presentation to the Business Matters Seminar, School of Business, Auckland University, Auckland, February 27, 2014. http://video.com.auckland.ac.nz/Video/Energy-Series/Tom-Murphy.mp4.

Mutu, Margaret. "Ngāti Kahu Kaitiakitanga meets Statoil and Shanghai CRED." Presentation to the Symposium on Māori Engagement with Extractive Industry, Faculty of Law, Auckland University, June 12, 2015.

New Zealand Institute of Economic Research. "East Coast Oil and Gas Development Study: Economic Potential of Oil and Gas Development." Report to the

Ministry of Business, Innovation and Employment, Wellington, November 2012. http://www.mbie.govt.nz/info-services/sectors-industries/natural-resources/oil-and -gas/petroleum-expert-reports/east-coast-oil-and-gas-development-study/NZIER -report-East-Coast-study-Value-oil-gas-exploration.pdf.

O'Brien, Thomas. "Fires and Flotillas: Opposition to Offshore Oil Exploration in New Zealand." *Social Movement Studies: Journal of Social, Cultural and Political Protest* 12, 2 (2013): 221–226.

Opalka, Katia. "Oil and Gas: the View from Canada." *Natural Resources & Environment* 26, 2 (Fall 2011).

Oreskes, Naomi, and Erik Conway. *Merchants of Doubt: How a Handful of Scientists Obscured the Truth on Issues from Tobacco Smoke to Global Warming.* New York: Bloomsbury Press, 2010.

Organisation for Economic Cooperation and Development. *Inventory of Estimated Budgetary Support and Tax Expenditures for Fossil Fuels.* OECD, Paris, 2011. http://www.oecd.org/site/tadffss/48805150.pdf.

Osborn, Stephen, Avner Vengosh, Nathaniel Warner, and Robert Jackson. "Methane Contamination of Drinking Water Accompanying Gas-well Drilling and Hydraulic Fracturing." *Proceedings of the National Academy of Sciences* 108, 20 (May 17, 2011): 8172–8176.

Owen, John, and Deanna Kemp. "Social License and Mining: A Critical Perspective." *Resources Policy* 38, 1 (2013): 29–35.

Palmer, Sir Geoffrey QC. "Protecting New Zealand's Environment: An Analysis of the Government's Proposed Freshwater Management and Resource Management Act 1991 Reforms." Report Commissioned by Fish and Game NZ, Wellington, September 11, 2013. Accessed July 21, 2014. http://www.fishandgame .org.nz/sites/default/files/Fish%20and%20Game%20RMA%20Paper%20-%20 FINAL%20PRINT.pdf.

Palmer, Sir Geoffrey. "Protecting New Zealand's Environment: An Analysis of Government's Proposed Freshwater Management and Resource Management Act 1991 Reforms." Commissioned report to Fish and Game New Zealand, Wellington. September, 2013. http://www.forestandbird.org.nz/files/file/RMA_Campaign/ FishGame_GeoffreyPalmer_092013.pdf.

Palmer, Sir Geoffrey. "Memorandum: Implications of King Salmon and Ruataniwha Dam Decisions." Commissioned report to Fish and Game New Zealand, Wellington. May 2014. http://www.fishandgame.org.nz/sites/default/files/ Palmer%20revised%20opinion%20-%20RMA%20Pt2%2C%20post%20King%20 Salmon%2C%20Ruataniwha%20-%20May%2714.pdf.

Parliamentary Commissioner for the Environment. "Creating Our Future: Sustainable Development for New Zealand, Government Strategies." Background paper, June 2002, New Zealand Government, Wellington. http://www.smartgrowthbop.org.nz/ media/1345/Creating-Our-Future-June-2002.pdf.

Parliamentary Commissioner for the Environment. "Evaluating the Environmental Impacts of Fracking in New Zealand: An Interim Report." New Zealand Government, Wellington, November 2012.

Parliamentary Commissioner for the Environment. "Drilling for Oil and Gas in New Zealand: Environmental Oversight and Regulation." New Zealand Government, Wellington, June 2014.

Parliamentary Commissioner for the Environment. "New Zealand's Contribution to the New International Climate Change Agreement." Submission to the Minister for Climate Change Issues and the Minister for the Environment. New Zealand Government, Wellington. June 3, 2015. http://www.pce.parliament.nz/.

PEPANZ. "Our Petroleum Future." Chairman Chris Bush's address to the 1st New Zealand Petroleum Summit, Wellington, September 19, 2012. Accessed from www.pepanz.co.nz, April 20, 2013.

Richardson, David, and Richard Denniss. "Mining the Truth: The Rhetoric and Reality of the Commodities Boom." Paper 7, Canberra: The Australian Institute (September, 2011).

Ridlington, Elizabeth, and John Rumpler. *Fracking by the Numbers: Key Impacts of Dirty Drilling at the State and National Level.* Environment America Research and Policy Center, Boston, 2013. Accessed November 9, 2013. http://www.environment america.org/sites/environment/files/reports/EA_FrackingNumbers_scrn.pdf.

Robinson, David. "NZ Petroleum Summit Launches into its First Day." PEPANZ Media release, Wellington, 2012. Accessed April 20, 2013. http://www.pepanz .com/assets/Uploads/PR-Summit-4.pdf.

Robinson, David. "Assessing the Impacts of Coming Developments." Paper to the conference on "Assessing the Impacts of Petroleum and Mineral Extraction in New Zealand." New Zealand Association of Impact Assessment, Wellington, December 11, 2012. Accessed April 18, 2014, https://www.nzaia.org.nz/uploads/ 1/2/3/3/12339018/nzaia_2012.pdf.

Robinson, David. "The Value of the Oil and Gas Industry to New Zealand." Presentation to the 2013 Advantage New Zealand conference, Auckland. April 28–May 1, 2013.

Rosenberg, Andrew A., Pallavi Phartiyal, Gretchen Goldman, and Lewis M. Branscomb. "Exposing Fracking to Sunlight: the Public Needs Access to Reliable Information about the Effects of Unconventional Oil and Gas Development in Order for it to Trust that Local Communities' Concerns Won't Be Ignored in Favor of National and Global Interests." *Issues in Science and Technology* 31, 1 (Fall 2014). http://issues.org/31-1/exposing-fracking-to-sunlight/.

Ross, Michael. "Does Oil Hinder Democracy?" *World Politics* 53, 3 (April, 2001): 325–361.

Royal Society of New Zealand. *Facing the Future: Towards a Green Economy in New Zealand.* Wellington, 2014. http://www.royalsociety.org.nz/expert-advice/ papers/yr2014/greeneconomy/.

Royal Society of New Zealand. *Transition to a Low-Carbon Future for New Zealand.* Wellington, April 2016. http://www.royalsociety.org.nz/media/2016/04/Report -Transition-to-Low-Carbon-Economy-for-NZ-1.pdf.

Rudel, Thomas, J. Timmons Roberts, and JoAnn Carmin. "Political Economy of the Environment." *Annual Review of Sociology* 37 (August, 2011): 221-238.

Rudzitis, Gundars, and Kenton Bird. "The Myth and Reality of Sustainable New Zealand: Mining in a Pristine Land." *Environment Magazine*, September/October, 2011. http://www.environmentmagazine.org/Archives/Back%20Issues/2011/November-December%202011/Myths-full.html.

SANZ. *Strong Sustainability for New Zealand: Principles and Scenarios.* Sustainability Aotearoa New Zealand. Auckland, 2009. Accessed November 7, 2014. www.phase2.org

Scambary, Benefict. *My Country, Mine Country: Indigenous People, Mining and Development Contestation in Remote Australia.* CAEPR Research Monograph 33. Canberra: ANU Press, 2013.

Schuller, Tisha. "Deescalating Through the Courts?" Colorado Oil and Gas Association CEO's blog, November 18, 2014. Accessed April 15, 2015. www.coga.org.

Schweitzer, Donald. "The Myth of Purifying Fracking Water in Saudi America: The Competition between Food, Drink and Energy Needs." *Truthout*, January 29, 2013. Accessed February 19, 2013. http://www.truth-out.org/opinion/item/14113-the-myth-of-purifying-fracking-water-in-saudi-america-the-competition-between-food-drink-and-energy.

Shauk, Zain, and Bradley Olson. "Colorado Drillers Show Sensitive Side to Woo Fracking Foes." *Bloomberg News*, August 28, 2014.

Shonkoff, Seth, Jake Hays, and Madelon Finkel. "Environmental Public Health Dimensions of Shale and Tight Gas Development." *Environmental Health Perspectives* 122, 8 (August 2014): 787–795.

Society of Petroleum Engineers International. "SPE Petroleum Resources Management System Guide for Non-Technical Users." Dallas, Texas (2007). www.spe.org.

Srocco, SR. "CONDITION RED: Fracking Shale Is Destroying Oil & Gas Companies Balance Sheets." SRSrocco Report, August 13, 2014. Accessed August 27, 2014. https://srsroccoreport.com/condition-red-fracking-is-destroying-oil-gas-companies-balance-sheets/.

Sustainable Business Council. *Vision 2050: The Report—Business as Usual is not an Option*, 2012, Auckland. http://biznz-sbc.squiz.net.nz/__data/assets/pdf_file/0015/57120/Vision-2050-NZ-Report-2012.pdf.

Sustainability Aotearoa New Zealand. *Strong Sustainability for New Zealand: Principles and Scenarios.* SANZ, Auckland, 2009. http://www.earthslimits.org/resources/2014/5/24/strong-sustainability-for-new-zealand.

Taranaki Regional Council. "Hydrogeologic Risk Assessment of Hydraulic Fracturing for Gas Recovery in the Taranaki Region". Report to Council February 17, 2012. Accessed April 20, 2013. http://www.trc.govt.nz/assets/Publications/guidelines-procedures-and-publications/hydraulic-fracturing/hf-may2012-graph-p19.pdf.

Technical Advisory Group. "Report of the Minister for the Environment's Resource Management Act 1991 Principles Technical Advisory Group." Ministry for the Environment, New Zealand Government, Wellington, February 2012. https://www.mfe.govt.nz/sites/default/files/tag-rma-section6-7.pdf.

Texas Sharon, Report on the conference *Media and Stakeholder Relations Hydraulic Fracturing Initiative 2011,* Houston, Texas, October 31–November 1, 2011. *Bluedaze* (blog). http//dl.dropbox.com/u/18503846/gashole conference/agenda .docx.

Union of Concerned Scientists. *Toward an Evidence-based Fracking Debate,* Full Report. Cambridge, MA, 2013. www.uscusa.org/HFreport.

Union of Concerned Scientists. *Science, Democracy and Fracking: A Guide for Community Residents and Policy Makers Facing Decisions over Hydraulic Fracturing.* Cambridge, MA, 2013. Accessed November 14, 2013. http://www.ucsusa .org/sites/default/files/legacy/assets/documents/center-for-science-and-democracy/ fracking-informational-toolkit.pdf.

United Nations World Water Development Report, Vol 1 *Water and Energy.* UNESCO, Paris, 2014. Accessed March 20, 2015. http://unesdoc.unesco.org/ images/0022/002257/225741E.pdf.

Urbina, Ian. "Drilling Down: Regulation Lax as Gas Wells' Tainted Water Hits Rivers." *New York Times,* February 26, 2011.

van der Elst, Nicholas, Heather Savage, Katie Keranen, and Geoffrey Abers. "Enhanced Remote Earthquake Triggering at Fluid-Injection Sites in the Midwestern United States." *Science* 341, 6142 (July 12, 2013): 164–167. doi: 10.1126/science.1238948.

van Gelder, Sarah. "Why Canada's Indigenous Uprising is About All of Us." *Yes Magazine,* February 7, 2013.

van Gelder, Sarah. "Why the Corporate Media's Climate Change Censorship Is Only Half the Story." *YES Magazine,* October 11, 2013.

Welker, Marina. *Enacting the Corporation: An American Mining Firm in Post-Authoritarian Indonesia.* Oakland: University of California Press, 2014.

Widener, Patricia. "Benefits and Burdens of Transnational Campaigns: A Comparison of Four Oil Struggles in Ecuador." *Mobilization: An International Quarterly* 12, 1 (2007): 21–36.

Widener, Patricia. "Global Links and Environmental Flows: Oil Disputes in Ecuador." *Global Environmental Politics* 9, 1 (February 2009): 31–57.

Widener, Patricia. *Oil Injustice: Resisting and Conceding a Pipeline in Ecuador.* London: Rowman & Littlefield, 2011.

Widener, Patricia. "A Protracted Age of Oil: Pipelines, Refineries and Quiet Conflict." *Local Environment* 18, 7 (2013): 841–842. http://dx.doi.org/10.1080/1354 9839.2012.738655.

Widener, Patricia. "Community Responses to the Social, Economic and Environmental Impacts of Oil Disasters and Natural Resource Extraction." Presentation to a public seminar on issues around oil and gas development, Gisborne District Council chambers, Gisborne, New Zealand, October 16, 2013.

Wilber, Tom. *Under the Surface: Fracking, Fortunes and the Fate of the Marcellus Shale.* Ithaca, New York: Cornell University Press, 2012.

World Wildlife Fund. *Fossil Fuel Finance in New Zealand, Part 1: Government Support.* Wellington, 2013. Accessed October 10, 2013. http://awsassets.wwfnz.panda .org/downloads/wwf_fossil_fuel_finance_nz_subsidies_report.pdf.

Zeese, Kevin, and Margaret Flowers. "US Climate Bomb is Ticking: What the Gas Industry Doesn't Want You to Know." *Truthout,* March 6, 2013. Accessed March 17, 2013. http://www.truth-out.org/news/item/14958-us-climate-bomb-is-ticking-what-the-gas-industry-doesnt-want-you-to-know.

Index

Pages references for figures are italicized.

Index

About the Author

Terrence M. Loomis holds an MA (first-class honors) in social anthropology from Auckland University and a PhD in economic anthropology from the University of Adelaide. He has over fifteen years of research and development consulting experience in the United States, Canada, Australia, Southeast Asia, the Pacific, and New Zealand. He was director of economic development for the Mdewakanton Dakota tribe of Prairie Island, Minnesota for four years and holds a certificate in economic development finance from the National Development Council of America. From 1997 to 2000, he was foundation professor of development studies in the School of Māori and Pacific Development at Waikato University, before becoming a social sector policy advisor with the New Zealand government. He is currently an independent researcher and a visiting research scholar in the Institute for Governance and Policy Studies at Victoria University.

Lightning Source UK Ltd.
Milton Keynes UK
UKOW03n1845161216
290203UK00001B/49/P